Phosphorus Chemistry in Everyday Living

Arthur D. F. Toy

Director, Eastern Research Center

Stauffer Chemical Co.

AMERICAN CHEMICAL SOCIETY

WASHINGTON, D. C.   1976

Library of Congress CIP Data

Toy, Arthur Dock Fon, 1915-
  Phosphorus chemistry in everyday living.

  Includes index.

  1. Phosphorus. 2. Phosphates. 3. Organophosphorus
compounds.

  I. Title.

QD181.P1T69   546'.712   75-44049
ISBN 0-8412-0293-1

To: HUI-I

CONTENTS

PREFACE

Much of the content of this book is based on lectures I gave to chemistry students at various colleges in connection with the Visiting Scientists Program sponsored by the American Chemical Society, the Manufacturing Chemists Association, and the Industrial Research Institute. Chemistry teachers have commented that they found the material to be helpful as a supplement to their regular textbooks and that it also gave them a better understanding of industrial research.

Some of the materials have also been the topics of talks I have given at local sections of the American Chemical Society and at industrial research laboratories. Practicing chemists who are not specialists in phosphorus chemistry, chemical executives, and chemical sales people have found that the subject matter provided them with a better appreciation of the importance of the roles phosphorus chemistry plays in their everyday living.

With a few exceptions, the arrangement of the book is based on the specific applications for various phosphorus compounds. This arrangement should be convenient for practicing chemists who are interested in finding various phosphorus compounds for the applications they have in mind.

Whenever possible, I include the "why" behind the development of various applications. The science of chemistry has now advanced sufficiently so that if the properties of a compound are known, some of its applications may be predicted. Many important uses, however, are still the result of accidental discoveries. In some cases, from the vantage point of hindsight, the original premise for a discovery appears downright illogical. Credit should be given to the many inventors who are keen enough to observe the unexpected and are able to take advantage of it to make important contributions.

I use formulas and chemical equations liberally. They are not hieroglyphics invented by chemists to scare people away from chemistry. Readers should regard them as pictorial representations of the descriptive text. They are used to make the text clearer. Toward that end, wherever I introduce a new chemical name in the text, I have tried to include its formula in parentheses next to it for the first few times. These formulas are also used in the chemical equations. A minimum of effort is needed, therefore, to correlate the text with the equations.

Acknowledgment

I want to thank my friends in the chemical industry and colleagues at Stauffer Chemical Co. for supplying me with specific items of information. I am indebted to E. C. Galloway, J. W. Kettle, and Robert Parry who generously spent valuable hours in reviewing the entire original manuscript. Their advice and suggestions are appreciated. My thanks to Susan Osborne and the late Barbara Laughlin for typing the manuscript and other assistance.

Stauffer Chemical Co.
Dobbs Ferry, N. Y.
August 1975

ARTHUR D. F. TOY

INTRODUCTION

Phosphorus is an unusual chemical element. While the element itself can burst into flame spontaneously, some of its compounds impart non-flammability to a host of materials. In fact, they participate in some of the best flame-retarding agents known.

From a biological standpoint, too, phosphorus shows a peculiar ambivalence. Tiny quantities are essential for the proper nutrition of plants, animals, and man. Life's reproductive processes depend on the phosphorus-containing genetic code carriers DNA and RNA. Also, life cannot go on without energy exchanges mediated by vital phosphates such as adenosine triphosphate—called ATP by biochemists. Yet some phosphorus compounds, such as the nerve gases, are so poisonous that small amounts are lethal.

My own introduction to this element came with a high-school chemistry demonstration. Our teacher added a few drops of a solution to a piece of filter paper, then set it aside, and went on with his lecture. Within a few minutes, much to our surprise, we noticed that paper ignited spontaneously. Our teacher explained that he had dissolved a little phosphorus in the solvent, carbon disulfide (CS_2), which he added to the paper. When the volatile carbon disulfide evaporated, it left behind the elemental phosphorus which quickly revealed its fiery nature by combining with oxygen in the air to produce the hot flame which ignited the filter paper.

This incident stayed with me vividly because when the teacher stepped out of the room, one of the students took a piece of filter paper and added quite a bit of the carbon disulfide–phosphorus solution to it. He wrapped this carefully with more paper and put it in his hip pocket to take home. Despite these precautions, the carbon disulfide must have evaporated because the phosphorus paper caught fire in his pocket, and he could not sit comfortably for quite a spell.

Within a single class of phosphorus compounds some uses are quite similar while others vary greatly. Take the case of the family of calcium phosphate compounds. Monocalcium phosphate (CaH_2PO_4) is used as a leavening acid in baking to make tender biscuits. Dicalcium phosphate ($CaHPO_4 \cdot 2H_2O$) is used as a polishing agent in toothpaste. Tricalcium phosphate [$Ca_5(PO_4)_3OH$] is the conditioning agent in salt which keeps it flowing freely during humid weather.

Even a single phosphorus compound can have an amazing range of applications—*e.g.*, sodium tripolyphosphate ($Na_5P_3O_{10}$). This is an important ingredient in household detergents. In fact it constitutes up to 45% by weight of most detergents displayed in supermarkets. Sodium tripolyphosphate is also used in applications quite unrelated to detergents. For example, one of the major ingredients used to pickle hams is sodium tripolyphosphate.

The tart taste of most carbonated cola beverages and root beer comes from their phosphoric acid content. Rustproofing of steel also involves treatment with phosphoric acid. A final example is organic phosphates which are used to make plastics pliable and workable and are also added to gasoline to make auto engines run more smoothly.

In short, phosphorus and its compounds show remarkable versatility. As shown in the main body of the book, phosphorus and its compounds are very important to our everyday living.

Phosphorus Chemistry in Everyday Living

The Discovery of Phosphorus (by William Pether after the painting by Joseph Wright, published Dec. 1, 1775). The discovery of phosphorus by Brandt, an alchemist, in 1669, is one of the many contributions to systematic chemistry that came out of the so-called "Alchemical Era." Alchemists added substantially over the years to the advancement of science, even though many of their discoveries were accidentally hit upon while pursuing the transmutation of metals or the universal medicine. Discoveries include arsenic, antimony, and bismuth. The "flash lighting" of the scene here shown reveals remarkable craftsmanship. Phosphorus, an essential constituent of protoplasm, nervous tissue, and bones, exists in three allotropic forms: white, black, and red. When white phosphorus is exposed to the air in the dark, it emits a greenish light, gives off white fumes.

1

Phosphorus the Element

More than 300 years ago, in 1669, Hennig Brandt, a Hamburg alchemist, like most chemists of his day was trying to make gold. He let urine stand for days in a tub until it putrified. Then he boiled it down to a paste, heated this paste to a high temperature, and drew the vapors into water where they could condense—to gold. To his surprise and disappointment, however, he obtained instead a white, waxy substance that glowed in the dark.

Brandt had discovered phosphorus, the first element discovered by man other than the metals and non-metals, such as gold, lead, and sulfur, that were known in ancient civilizations. The word phosphorus, comes from the Greek and means light bearer.

Although news of the discovery spread quickly throughout Germany, Brandt remained secretive. He sold his method first to Dr. Johann Daniel Krafft of Dresden and later to others. Krafft, in turn, tried to sell it and traveled through Europe and eventually to America exhibiting and demonstrating this luminous substance. A few other chemists learned how to make phosphorus after making deals with Brandt or Krafft or by working out their own method from hints of the original one. Robert Boyle, the illustrious English scientist, is said to have discovered the process independently. However, it was not until 1737 when the French Government studied the process and published its report that this carefully guarded secret became public knowledge (1).

Pure phosphorus is a white, transparent, crystalline, waxy solid which glows in the dark and ignites in air just above room temperature (30°C). Most of the commercially produced element, however, is not pure; it is yellowish and semi-transparent. Hence the terms yellow and white phosphorus are used for the same material. It melts at 44.2°C but is a solid at room temperature and can easily be cut with a knife. Since it ignites readily when exposed to air, it is usually stored under water or in a closed container under a layer of inert gas such as carbon dioxide.

Alchemist Brandt made his phosphorus, which he called "cold fire," in a single, small pot, but today, phosphorus manufacture is a huge industry. In 1974 the United States alone produced more than a billion pounds of phosphorus. Today phosphorus is made in huge electric furnaces rather than in a small pot over a charcoal fire. In spite of this difference of scale, the chemistry of the two processes is remarkably similar. Brandt's process was based on a high temperature reaction of inorganic phosphates with carbon. Inorganic phosphates, such as calcium phosphates and magnesium ammonium phosphates, are present in urine, and carbon is present in the form of organic compounds. The modern process for making phosphorus has improved Brandt's method only to the extent of adding silica—sand—to the reaction mixture.

Mining

For some time, calcium phosphate from animal bones was used as the source of inorganic phosphate, but eventually this supply became inadequate. By the early 1800's, the need for bones in England became so great that European battlefields were being combed for them. Present sources of calcium phosphate are minerals, commonly known as phosphorite or phosphate rocks. They are mined in enormous, open pits with drag line buckets so large that an automobile can turn around in them. The phosphate rocks have an approximate composition of the mineral fluorapatite $[Ca_5(PO_4)_3F]$, a calcium phosphate containing 4% fluorine. They are contaminated with iron and aluminum oxides, carbonates, silicates, and organic matter. The largest known American deposits of phosphate rocks lie in Florida, Tennessee, North and South Carolina, Kentucky, Virginia, and the western states of Utah, Idaho, and Montana. In other parts of the world, large deposits are found in the Kola Peninsula, near Kirovsk in the Soviet Arctic part of Scandinavia. Important deposits are also located in Algeria, Tunisia, Egypt, and Morocco. The origin of these phosphate rocks is believed to be the small quantities of phosphate normally present in earth's granite rocks. Over eons, through weathering and leaching, they eventually found their way into the sea where marine animals absorbed and concentrated them into their shells, bones, and tissues. The remains of these animals accumulated on the bottom of the sea, and over the years, through geological changes, some of the phosphatic sediments appeared as deposits in dry land.

Electric furnaces for producing phosphorus are usually located near phosphate deposits and a source of cheap electric power. In the United States most phosphorus furnaces are in Florida, Tennessee, Utah, and Montana.

Manufacture

Mined phosphate rocks are actually fine granular particles. Before charging into the electric furnace, they are first washed to remove most of the clay impurities. They are then fused in a high temperature kiln into nodules about an inch in diameter, matching in size the pieces of silica and coke co-reactants. Coke is the source of carbon for the reaction.

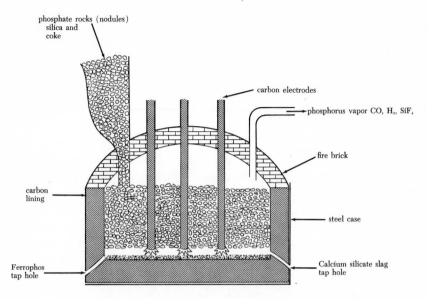

Figure 1. Schematic of electric phosphorus furnace

After the phosphate nodules, silica, and coke are loaded into the furnace (*see* Figure 1), a high voltage electric arc is struck in the furnace to produce temperatures of 1200°–1450°C.

The reaction is still not clear; chemists have speculated on it for years. Most textbooks show it in one step:

$$2Ca_3(PO_4)_2 + 6\ SiO_2 + 10\ C \longrightarrow 6\ CaSiO_3 + 10\ CO + P_4$$

Another view is that it takes place in two steps. When the mixture is heated to a high temperature, silica unites with the calcium oxide portion of the calcium phosphate to liberate phosphorus pentoxide. When the temperature reaches 1450°C in the electric furnace, the phosphorus oxide vaporizes. As this vapor meets carbon, it is reduced to elemental phosphorus while the carbon is oxidized to carbon monoxide. In the two reactions here calcium phosphate is written in the alternative form of two oxides:

$$2(3CaO \cdot P_2O_5) + 6\ SiO_2 \longrightarrow 2\ P_2O_5 + 6\ CaSiO_3$$
$$2\ P_2O_5 + 10\ C \longrightarrow P_4 + 10\ CO$$

Still another theory visualizes three reactions. First the carbon reacts with calcium phosphate to form carbon monoxide (CO) and calcium phosphide (Ca_3P_2). The calcium phosphide then reacts with more calcium phosphate to form phosphorus and by-product calcium oxide. The calcium oxide reacts immediately with silica (a base reacting with an acid) to form the salt, calcium silicate:

$$Ca_3(PO_4)_2 + 8\ C \longrightarrow Ca_3P_2 + 8\ CO$$
$$Ca_3P_2 + Ca_3(PO_4)_2 + 2\ C \longrightarrow P_4 + 2\ CO + 6\ CaO$$
$$6\ CaO + 6\ SiO_2 \longrightarrow 6\ CaSiO_3$$

The summation of these equations, and of those in the two-step theory, is identical to the one-step equation. There are other ways in which the reaction could take place, and chemists are still actively studying the mechanism of this phosphorus producing reaction. With a better understanding of the nature of the reaction, it may be possible to improve the process further and to obtain a higher phosphorus yield.

Continental Oil's Agrico Chemical Division elemental phosphorus plant at Pierce, Fla. Flames are burning excess carbon monoxide.

Silica in the furnace mixture is vital to large-scale production of phosphorus. It removes the calcium oxide that is formed so that the reaction can continue. Calcium oxide requires a very high temperature to melt. At the 1450°C inside the electric furnace it is still a solid and would be difficult to extract in order to recharge the furnace with more raw material. Silica reacts with calcium oxide to form calcium silicate, which melts at furnace temperature and collects at the bottom where it can be easily drained away in a spectacular ribbon of fire.

High pressure water hoses break down tons of phosphate rock and float it to the processing plant at this Bartow, Fla. installation of International Minerals & Chemical Corp. The raw ore is cumped into place by a huge dragline, in the background. The bucket on the largest IMC dragline can hold 75 tons of ore.

Other chemicals are formed in the modern phosphorus furnace since phosphate rock also contains iron oxide, calcium fluoride, and clay. Even though the rock is washed to remove most of the clay, it is still quite impure.

Silica in the furnace mixture reacts with the calcium fluoride impurity to form silicon tetrafluoride, a toxic compound that is a gas at furnace temperature. To avoid polluting the atmosphere with this gas, the furnace vapors are scrubbed with a lime solution.

The iron oxide impurity is taken care of easily. High temperature carbon reduces it to molten iron, which in turn reacts with phosphorus to form iron phosphide, commonly called ferrophos. Ferrophos is a heavy liquid at electric furnace temperatures. As it forms, it sinks to the bottom of the furnace beneath the molten calcium silicate and can be drained away. Only then is the molten calcium silicate removed.

To the phosphorus producer, formation of by-product ferrophos is a nuisance. Every 100 lbs of ferrophos formed takes about 25 lbs of his precious phosphorus. This is a major loss, but an unavoidable one since the iron oxide is present in phosphate ore and cannot be removed beforehand economically. Today, the few uses of ferrophos depend on its heavy density. As a fine powder, it is mixed with dynamite for blasting. Its high density also makes it useful as a filler in high density concrete used as radiation shields in nuclear reactor installations.

Phosphorus producers sell the calcium silicate slag by-product as ballast for roadbeds in highway and railway construction.

Some efforts have also been made to recover fluorine from the by-product silicon tetrafluoride. Reaction with water and soda ash converts silicon tetrafluoride to sodium silicofluoride (Na_2SiF_6), which is used for the fluoridation of water.

The carbon monoxide by-product from the phosphorus producing reaction is also of value. Since it burns in air to form carbon dioxide and release heat, it is used as fuel to supplement the heat needed to fuse the calcium phosphate mineral into nodules needed for the furnace reaction.

As a result of all these assorted manipulations, losses, and gains, the phosphorus produced sold for a little less than $400 a ton, or 20¢ a pound in 1972. Because of inflation and higher costs, for energy and phosphate rocks, the price for phosphorus in 1974 had gone up to above $1,000 a ton.

Literature Cited

1. Weeks, Mary E., "Discovery of the Elements," 4th ed., Mack Printing Co., Easton, Pa., 1939.

2

Phosphorus in Matches and in Warfare

Early chemists must have been intrigued by the fact that phosphorus ignites easily in air, especially since fire had to be started by nursing flame from a spark produced by striking steel and flint together. Although the idea of starting a fire with phosphorus occurred to many people, modern matches that we take so much for granted, were not developed overnight. More than 200 years passed before the familiar book matches of today evolved.

The early phosphorus matches were crude. In the 1780's, the expensive "ethereal match" was fashionable. It consisted of a slip of paper tipped with white phosphorus and sealed in a glass tube. When a flame was needed, the glass tube was broken, air entered, oxidized the phosphorus, and ignited the paper. A modification of this device was the "phosphorus box" which consisted of sulfur-tipped wood splints sold with a small bottle of phosphorus. To start a fire, the splint was dipped in the phosphorus and then held in air. Air burned the phosphorus, which burned the sulfur, which burned the splint.

The next improvement was the "strike-anywhere" match. Research showed that white phosphorus does not ignite as readily when coated with glue or starch. It will ignite, however, by heat generated from friction such as by striking. This discovery led to the development of today's wooden matches. The strike-anywhere match consists of a head of phosphorus as the igniting agent and underneath a body of potassium chlorate ($KClO_3$) as the oxidizing agent. They are held together by animal glue and starch around the sulfur-coated tip of a wood splinter. This invention takes advantage of a special property of potassium chlorate, which loses all its oxygen when heated or rubbed with phosphorus, by oxidizing the phosphorus to phosphorus oxide (P_4O_{10}). The reaction generates large amounts of heat. When a phosphorus-tipped chlorate match is struck, frictional heat causes the phosphorus to ignite in air. At the same time, the phosphorus and organic binders are rapidly oxidized by potassium chlorate. The heat of the reaction quickly ignites the sulfur which transmits the flame to the wooden splint. Since this rapid reaction can be explosive, inert materials, such as diatomaceous earth or ground glass, are added to retard combustion.

We face two major problems in using white phosphorus for matches. First, it is extremely poisonous and can cause "phossy jaw," an illness which exacted a fearful toll among early matchmakers. The disease is so named because it is caused by phosphorus fumes that are inhaled and also enter the body through decayed teeth to attack and destroy bones, particularly jaw bones. "Phossy jaw" is usually fatal and was very prevalent before automatic matchmaking machines were invented. Since matches were first made by hand in the homes of the poor, many of them suffered this horrible death. Other innocent persons, including babies, were poisoned by accidentally eating or chewing yellow phosphorus matches. The fatal dose of phosphorus to humans is about one-tenth of 1 gram, less than 1/100 of an ounce. ("Phossy jaw" is a constant threat even today to people who work with elemental phosphorus. The danger, however, is minimized by avoiding phosphorus vapors and by strict adherence to good dental care—absolutely no unfilled cavities.)

The second problem encountered with white phosphorus matches is that they ignite so easily. Accidental fires can be readily started, sometimes even by the gnawing of rats and mice. Obviously white phosphorus had to be replaced with a material which would be safer—both health- and fire-wise.

This material turned out to be a non-crystalline (amorphous) form of the element called red phosphorus. It is produced by heating white crystalline phosphorus to about 400°C under an atmosphere in which the active oxygen has been replaced by a more inert gas such as nitrogen or argon. During heating, white phosphorus (P_4), which has a tetrahedral molecular structure containing four phosphorus atoms:

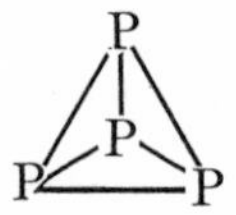

is believed to rearrange by the cleavage of one of the P–P bonds. The cleaved bonds of many P_4 molecules then join to form a polymer of the following chain structure:

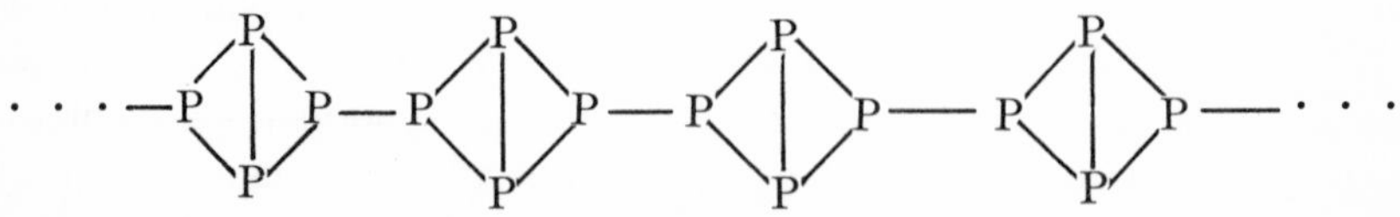

Red phosphorus with this structure is not poisonous and has an ignition temperature nine times higher than that of the white form (260° vs. 30°C). The development of red phosphorus paved the way for safety matches. Despite its higher ignition temperature, amorphous red phosphorus can still speed the rapid oxidative action of potassium chlorate ($KClO_3$).

Safety matches (wooden as well as book matches) have as basic ingredients a head of potassium chlorate and sulfur bound with glue around the tip of a splint that has been dipped in paraffin. The separate friction striking surface has two basic ingredients—mostly red phosphorus as igniting agent and glass powder to provide the friction. These components are bound by glue. Zinc oxide and calcium carbonate stabilizers are added to protect the red phosphorus against exposure to air and moisture. Without such stabilizers, the striking surface would turn black and soggy and would not light the match when struck.

The wooden or paper splint of the match is treated first with a solution of monoammonium phosphate ($NH_4H_2PO_4$) to prevent afterglow when the flame is extinguished. The no-afterglow match is a modification of the original match in which the splint was treated chemically so that it would not burn beyond the mid-point. For obvious reasons, that item was called the drunkard's match. The no-afterglow treatment has prevented many accidental fires from careless users who discard a still smouldering match. Sulfur was once used with phosphorus in strike-anywhere and safety matches. However, at the end of the 19th century phosphorus sesquisulfide (P_4S_3), in which phosphorus and sulfur are chemically combined in a 4P to 3S ratio, was introduced as the igniting agent. This compound is relatively non-toxic and is unaffected by the atmosphere. It does not react with water at ordinary temperatures. Above 100°C, it burns in air. Produced by reaction of white phosphorus with molten sulfur at 320°–380°C, it is purified of other phosphorus sulfide impurities having P to S ratios, such as 4 to 10 and 4 to 7, by washing with water or dilute bicarbonate of soda and drying at 40°–50°C.

The ingredients of our modern match are thus a tip of phosphorus sesquisulfide as igniting agent, with zinc oxide as stabilizer and potassium chlorate as the oxidizer, held by glue with powdered glass. The bulb contains less phosphorus sesquisulfide and more potassium chlorate to assist the initial oxidation whereby the match splint is raised above its flame point.

The next time you strike a match, consider all the chemical ingenuity that has gone into that small, instant flame. Although the invention and development of matches has solved many of the problems of obtaining an instant flame, man's quest for improved products is never ending. Phosphorus-containing matches have to be thrown away after each use. What a challenge it would be to develop a reusable permanent match other than the cigarette lighter.

Use of Phosphorus for Incendiaries

White phosphorus burns rapidly in air to form phosphorus oxide (P_4O_{10}). In the atmosphere, phosphorus oxide appears as a dense white smoke. During World War I, the military forces discovered that the obscuring power of this smoke per unit weight of phosphorus was greater than any smoke-generating chemicals then known. Taking advantage of this property, munitions makers produced many phosphorus shells for artillery use. The opaque cloud of phosphorus oxide smoke hides an army from the enemy, yet is sufficiently innocuous for soldiers to penetrate it with little discomfort. Phosphorus-filled shells and grenades continued to be used in World War II. In application on the battlefield, they were also effective weapons. A barrage of phosphorus shells rained small particles of burning phosphorus which stuck tenaciously to clothing and skin, burning the enemy to death horribly.

Phosphorus shell manufacture was refined in World War II. Normally the shells are filled with molten phosphorus which solidifies into a large chunk. Upon exploding, solid phosphorus is broken up and ignited, but this produces a non-uniform distribution of white phosphorus smoke for screening. If the solid phosphorus in small particles is uniformly mixed with a rubber–gasoline gel, the smoke screen obtained is more homogeneous.

Self-igniting phosphorus when combined with benzene is called a Molotov cocktail. During World War II, right after the evacuation from Dunkirk, millions of Molotov cocktails were made by the British with beer and milk bottles for the defense of England. Storage of these bottles gave the local citizenry rather exciting experiences. For safety, the bottles were usually submerged in a nearby stream. Occasionally, however, boxes of a thousand or so bottles broke loose and floated away, causing considerable concern and several spectacular fireworks displays.

The sensitivity of phosphorus to ignition is also used for amusement. Caps for toy pistols are made with a mixture of two separate pastes—one of potassium chlorate and the other of red phosphorus—sulfur and calcium carbonate. A water slurry of the pastes containing a gum binder is dabbed on paper to make caps. The impact of metal on the mixture causes the explosive ignition heard when the trigger is pulled.

3

Phosphoric Acids

A hot day and vigorous exercise will make most of us reach for a refreshing cola drink. It never occurs to us that we are actually drinking a flavored, carbonated, sweetened, dilute solution of phosphoric acid. We give even less thought to the fact that the bottle cap is made from sheet metal that has been phosphatized in a phosphoric acid solution so that it won't rust through the enamel coating; or that the bottle is cleaned in an alkaline bath containing sodium phosphates. Without realizing it, we use phosphoric acid and its derivatives every day of our lives.

Although the phosphoric acid family has many members, only two can be isolated in large quantities: orthophosphoric and pyrophosphoric acid. Orthophosphoric acid—or simply phosphoric acid—is manufactured by two processes—thermal and wet.

Thermal Phosphoric Acid

Thermal phosphoric acid is the general term applied to phosphoric acid prepared by the reaction of water with phosphoric anhydride (P_4O_{10}, more commonly called P_2O_5) obtained by burning elemental phosphorus in air. The reactions are:

$$P_4 + 5\ O_2 \longrightarrow P_4O_{10}$$

phosphorus + oxygen   phosphoric anhydride

$$P_4O_{10} + 6\ H_2O \longrightarrow 4\ H_3PO_4$$

phosphoric acid

Although this is a two-step process, in practice the reactions are carried out consecutively in a single system of reactors. The phosphoric anhydride (P_2O_5) formed by burning phosphorus is hydrated immediately with water. Both reactions generate considerable heat; hence the term "thermal acid."

The heat generated is removed by cool water, producing a great cloud of steam from the tops of the reactors. On a cold day it looks like the P_2O_5 smoke screen from a phosphorus bomb. At a production plant where I was once working, a pollution control officer thought we were allowing P_2O_5 to escape into the atmosphere. Our engineer had to climb with him to the top of the reactors, condense some of the steam, and then drink it to prove that the vapor was harmless.

The concentration of industrially produced phosphoric acid is usually 80–90%. It contains some arsenic as arsenic oxide or arsenic acid. The source of arsenic is calcium arsenate impurity in the phosphate rock reduced in the electric furnace to simple arsenic. It co-distills with phosphorus and oxidizes along with it during burning. Since a lot of thermal phosphoric acid is used in food, toxic arsenic must be removed. The usual technique is to add gaseous hydrogen sulfide that smells like rotten eggs. Arsenic is precipitated as the insoluble arsenic sulfides and is removed by filtration. Any residual hydrogen sulfide is then removed by blowing air through the acid.

Next, the treated phosphoric acid is diluted to 85, 80, or 75% strength—the common commercial concentrations. It can also be concentrated by evaporating the water at high temperature to give the equivalent of 105% H_3PO_4. This is "superphosphoric acid," and its P_2O_5 content is 76.5%. Finally, it can be concentrated to 82–84% P_2O_5, the strength of commercial polyphosphoric acid.

Wet Process Phosphoric Acid

Unlike thermal phosphoric acid, wet process phosphoric acid is made directly from calcium phosphate rock. Sulfuric acid reacts with calcium phosphate to form water-insoluble calcium sulfate ($CaSO_4 \cdot 2 H_2O$, or gypsum) and a water solution of phosphoric acid. The gypsum precipitate is removed by filtration, and the dilute water solution of phosphoric acid is then concentrated by boiling off water until the desired phosphoric acid concentration is obtained.

When the mineral fluorapatite, $Ca_5(PO_4)_3F$, is used as the calcium phosphate source, the reaction is:

$$Ca_5(PO_4)_3F + 5 H_2SO_4 + 10 H_2O \longrightarrow$$

$$\underset{\text{gypsum}}{5\ CaSO_4 \cdot 2 H_2O} \quad + \quad \underset{\substack{\text{phosphoric} \\ \text{acid}}}{3\ H_3PO_4} \quad + \quad \underset{\substack{\text{hydrogen} \\ \text{fluoride gas}}}{HF}$$

Commercial production of wet process phosphoric acid dates back to about 1850. The original commercial production of elemental phosphorus (on a very small scale) used wet process phosphoric acid as raw material. Most earlier wet process acid was used to manufacture monocalcium phosphate for use in baking powder. By the 1870's, the acid was being used in Germany to make fertilizers.

Originally, animal bones were the source of calcium phosphate. In England at one time the shortage of bones was so acute that old battlefields in Europe were turned to for their supply. In many areas bones were still the raw material into the 20th century.

August Kochs, the founder of Victor Chemical Works, later acquired by the Stauffer Chemical Co., was one of the pioneers in the manufacture of phosphoric acid and phosphates. He started his business near Chicago in the early 1900's. Every morning he would drive his horse-drawn wagon to the Chicago stockyard to buy animal bones. In the afternoon he went out to sell his products. For 50 years the phosphate industry grew, and so did his business. Animal bones were eventually replaced by phosphate minerals, and wet process phosphoric acid for food grade was superseded by thermal phosphoric acid. By the time Kochs died, the business he founded was sold for over 100 million dollars.

The greatest difficulty in manufacturing wet process phosphoric acid is to control the reaction conditions so that precipitated gypsum is easy to filter and wash free of residual phosphoric acid. Originally, only a 15% concentration of phosphoric acid could be made directly. Today, through process modifications, concentrations of 40–50% before evaporation are possible. Even though these concentrations are good enough for many applications, impurities such as fluorosilicate, iron, and alumi-

Wet process phosphoric acid plant in Pierce, Fla.

Thermal phosphoric acid plant. Note its compact size in relation to the wet process plant shown on page 13.

num must be removed to produce the industrial grade sodium phosphates used in detergents and cleaning products (but not in food).

Applications. Most wet process phosphoric acid is used in fertilizer. Fertilizer triple superphosphate is prepared by the action of wet phosphoric acid on finely ground phosphate rock. The wet process acid—when concentrated to superphosphoric acid—is used to make high analysis liquid fertilizer.

Industrial grade sodium phosphates are made after removing some impurities from wet process acid. First, the acid solution is adjusted with sodium carbonate (Na_2CO_3, or soda ash) to a pH of about 2. At this point most of the fluoride impurity precipitates out as insoluble sodium fluorosilicate (Na_2SiF_6) and filtered off. (Incidentally, pure sodium fluorosilicate is used to fluoridate water.)

Next the pH of the solution is raised to around 5. At this point most of the iron and aluminum impurities, along with some residual sodium fluorosilicate, come out. The mixture is filtered again, and the resulting solution is essentially monosodium phosphate. This compound can be converted to other sodium phosphates by reaction with more soda ash or sodium hydroxide or by other treatments. These phosphate salts still contain impurities that make them unsuitable for food applications, but they can be used in detergents and other cleaning products.

According to the U.S. Bureau of Census the total American production of phosphoric acid for 1973 was equivalent to 6,493,000 tons of 100% P_2O_5. Of this, about one-third was produced from elemental phosphorus.

New Process. Wet process phosphoric acid is made with sulfuric acid rather than hydrochloric acid so that the by-product, gypsum, which is insoluble in water can be removed by filtration. If hydrochloric acid were used (as shown below) the by-product would be water-soluble calcium chloride ($CaCl_2$):

$$Ca_5(PO_4)_3F + 10\ HCl \longrightarrow 3\ H_3PO_4 + 5\ CaCl_2 + HF$$

Separation of water-soluble calcium chloride from a water solution of phosphoric acid is a sticky problem, but it was solved neatly by Israel Mining Industries (IMI). Phosphate rock is digested with hydrochloric acid to give a water solution of phosphoric acid and calcium chloride. The acid is then selectively extracted using an organic solvent. Phosphoric acid is then stripped from the organic solvent with water. Both butyl and amyl alcohol are water insoluble and can dissolve phosphoric acid but not calcium chloride. Other solvents have been found suitable also.

Even crude phosphoric acid obtained by the sulfuric acid wet process can be purified by the solvent extraction described above. After each extraction stage, more and more impurities are left in the discarded water layer. The number of extractions used depends on the purity needed for the specific end use. Many problems remain in solvent extraction, and for now the quantity of phosphoric acid produced by this route is not significant.

Classes of Phosphoric Acids

There are many members in the phosphoric acid family. If they are examined individually, their relationship to each other will be clear. (Superphosphoric acid is discussed in Chapter 6 on fertilizers.)

The first member is orthophosphoric acid; it is common phosphoric acid and usually goes without the prefix ortho-. Its formula is H_3PO_4. If two molecules of phosphoric acid are heated to remove one molecule of water, the result is pyrophosphoric acid (also called diphosphoric acid) containing two phosphorus atoms. Its P_2O_5 content is 79.8%. The reaction is shown below (top of p. 16).

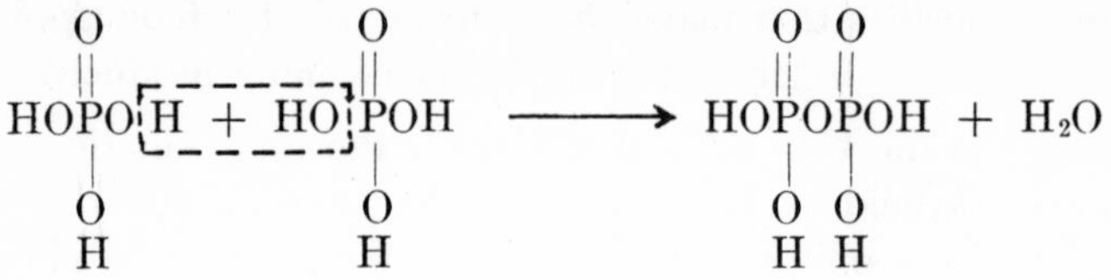

$$
\underset{\text{orthophosphoric acid}}{\text{HOPO─H + HO─POH}} \longrightarrow \underset{\text{pyrophosphoric acid}}{\text{HOPOPOH + H}_2\text{O}}
$$

If three molecules of phosphoric acid are heated to eliminate two molecules of water, tripolyphosphoric acid is formed:

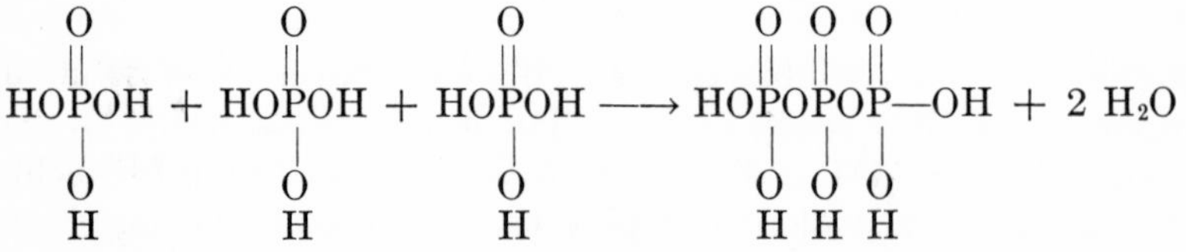

$$
\underset{\text{orthophosphoric acid}}{\text{HOPOH + HOPOH + HOPOH}} \longrightarrow \underset{\text{tripolyphosphoric acid}}{\text{HOPOPOP─OH + 2 H}_2\text{O}}
$$

Tripolyphosphoric acid has a P_2O_5 content of 82.6%. By continuing in this way, it is theoretically possible to make phosphoric acids of various chain length containing more and more phosphorus atoms. An acid with four phosphorus atoms is called tetrapolyphosphoric acid, one with five is called pentapolyphosphoric acid, and so on.

Three molecules of orthophosphoric acid can also be condensed into one molecule by eliminating three instead of two moles of water. Theoretically, in this case they could form a cyclic trimetaphosphoric acid:

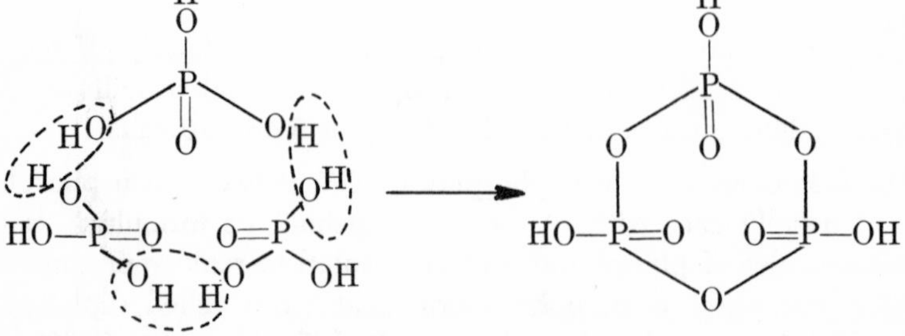

orthophosphoric acid   cyclic trimetaphosphoric acid

Similarly, four moles of phosphoric acid theoretically could condense into a cyclic tetrametaphosphoric acid by eliminating four moles of water.

Although these cyclic acids exist, they are not made by this water-elimination method. They are only mentioned to give a complete picture of phosphoric acids and polyphosphoric acids and their relationship to each other.

Properties of Phosphoric Acids. Except for orthophosphoric and pyrophosphoric acid, it is quite difficult to isolate in large quantities the individual polyphosphoric acids. As indicated earlier, when two moles of phosphoric acid are heated to remove one mole of water, the residue contains 79.8% P_2O_5. Since this is the exact theoretical percentage of P_2O_5 for pyrophosphoric acid, we would expect this to be pure pyrophosphoric acid. However, analysis by paper chromatography shows that it contains only about 43% pyrophosphoric acid; the remainder is about 17% ortho-, 23% tripoly-, 11% tetrapoly-, a lesser amount of pentapoly-, as "hypoly"-phosphoric acid.]

[Paper chromatography is a special technique for separating and detecting compounds. When compounds of the same class are treated with a proper solvent on absorbent paper such as filter paper, the smallest (lowest molecular weight) moves the fastest and the farthest away from the starting point. The largest (highest molecular weight) moves the slowest and travels the shortest distance. The in-between members travel varying distances in inverse relation to their molecular weights. Thus, when a mixture of phosphoric acid of various chain lengths or molecular weights is put as a spot on a filter paper and washed with solvent, after some time, orthophosphoric acid will have moved the farthest from the origin, followed by pyro-, then tripoly-, tetrapoly-, pentapoly-, and so on. Since the acids are colorless, these migration spots cannot be seen. However, they can be made visible if they are all converted to the orthophosphate by partial hydrolysis and reaction with special reagents of molybdate and stannous chloride to produce blue spots. Present techniques can separate polyphosphoric acids as large as those with nine phosphorus atoms. Polyphosphoric acids above nine are lumped together as "hypoly-" phosphoric acid.]

When three moles of phosphoric acid are condensed by eliminating two moles of water, the residue has a P_2O_5 content of 82.6%, the theoretical value for pure tripolyphosphoric acid. Paper chromatographic analysis shows a content of only about 17% tripolyphosphoric acid, the total being various amounts of acids of different chain lengths ranging from the ortho with one phosphorus atom to acids with two, three, four, five, six, seven, eight, nine phosphorus atoms on up.

Syrupy condensed phosphoric acid with a specific P_2O_5 content is thus not a pure compound. It is a mixture of polyphosphoric acids in

Table I. Composition of

Composition, wt % P_2O_5	*Ortho-*	*Pyro-*	*Tri-*	*Tetra-*
68.80	100.00	Trace		
69.81	97.85	2.15		
70.62	95.22	4.78		
72.04	89.91	10.09		
72.44	87.28	12.72		
73.43	76.69	23.31		
74.26	67.78	29.54	2.67	
75.14	55.81	38.88	5.31	
75.97	48.93	41.76	8.23	1.08
77.12	39.86	46.70	11.16	2.28
78.02	26.91	49.30	16.85	5.33
78.52	24.43	48.29	18.27	6.75
79.45	16.73	43.29	22.09	10.69
80.51	13.46	35.00	24.98	13.99
81.60	8.06	27.01	22.28	16.99
82.57	5.10	19.91	16.43	16.01
83.48	4.95	16.94	15.82	15.91
84.20	3.63	10.60	11.63	13.05
84.95	2.32	6.97	7.74	11.00
86.26	1.54	2.97	3.31	5.16

[a] Huhti, Anna-Liisa, Gartaganis, Phoebus A., *Canadian Journal of Chemistry* (1956) **34**, 790. Reproduced by permission of the National Research Council of Canada.

equilibrium with each other. The P_2O_5 content controls the equilibrium composition. The equilibrium composition of the acid is the same whether it is made by removing water from orthophosphoric acid or by adding P_2O_5.

Orthophosphoric acid has a theoretical P_2O_5 content of 72.4% (*see* Table I). According to some investigators, a liquid phosphoric acid with this P_2O_5 content contains less than 90% orthophosphoric acid, the remainder being pyrophosphoric acid and some unreacted water. How-

Strong Phosphoric Acids[a]

Penta-	Hexa-	Hepta-	Octa-	Nona-	"Hypoly-"
1.60					
2.26					
4.48	1.92	0.80			
6.58	3.14	2.84			
11.00	5.78	3.72	2.31	1.55	1.28
12.64	8.89	6.41	4.11	3.51	6.99
12.46	9.71	6.77	5.04	2.99	9.42
12.17	9.75	8.19	5.92	4.91	20.16
10.45	9.62	8.62	7.85	6.03	29.41
5.32	5.54	3.51	3.30	3.30	66.03

ever, when it is crystallized from a liquid of 72.4% P_2O_5 content, the equilibrium shifts toward the pure H_3PO_4 and eventually it is all converted to pure, solid crystalline orthophosphoric acid with a melting point of 42.35°C. Similarly, the pure pyrophosphoric acid with a melting point of 54.3°C can be crystallized in the pure form from a liquid mixture of phosphoric acids having a theoretical average P_2O_5 content of 79.8%. Ortho- and pyrophosphoric are the only acids which have been isolated in a pure form in this manner.

4

Food Uses of Phosphates

If a tooth is allowed to stand in a glass of carbonated cola drink overnight, some people say it will completely dissolve. When my son was
about five years old, I decided to test this theory. My wife and I had tried
to discourage him from drinking too much soda by telling him that the
dilute acid solution was not good for his teeth. But that explanation
didn't impress him. Hoping to demonstrate our point graphically, we
put a baby tooth he had just lost into a glass of well-known cola overnight, predicting that it would dissolve by morning. When I awoke first
and found the tooth still there, I fished it out. When our son saw that
the tooth was gone, he was astonished at this evidence of cola corrosion,
but as honest parents, we confessed the little trick we had played on
him. However, to show that we weren't entirely wrong, we urged him to
feel the tooth. It had become quite soft from the etching action of the
phosphoric acid.

For Tartness in Carbonated Beverages—Phosphoric Acid

The tartness which makes drinks so refreshing comes from added
acids, such as citric (citrus fruit acid), tartaric (grape acid), malic
(apple acid), and phosphoric. Fruit acids are used in fruit-flavored
drinks. Cola drinks, root beer, and sarsaparilla generally contain phosphoric acid in amounts ranging from 0.013 to 0.084% of 75% phosphoric
acid. Food grade phosphoric acid used in these drinks must be very
pure, and quality specifications are the most stringent in the business.
In analyzing for phosphoric acid content, the cola manufacturer requires
three different methods, all of whose results must cross-check.

The acidity of cola is measured by determining its pH, that is, the
negative logarithm of its hydrogen ion concentration. The higher the pH,
the lower the acidity, a pH of 7 being neutral. Cola contains from 0.057
to 0.083% of 75% phosphoric acid. It is strongly acidic, with a pH of
around 2.3 to 2.5. Root beer, which contains only about 0.013% of 75%
phosphoric acid, has a pH close to 5. The pH of the human stomach
normally runs around 2.5 and during digestion can go as low as 0.9.
Each acid, depending on its chemical nature, gives a different hydrogen
ion concentration when the same amount is dissolved in an equal volume
of water. In other words, to obtain solutions of the same acid strength,
different weights of the different acids are needed.

Acid strength is reflected in the sourness or tartness of a soft drink. The stronger the acid, the more sour the drink. Acid strength can also be influenced by other ingredients—*i.e.*, buffering agents. The only true test for sourness is the taste test. In a sample test, on a 100% basis, one pound of phosphoric acid was equal in sourness to 4.25 lbs. of citric acid or 5.5 lbs. of tartaric acid. Since citric acid costs about 30¢ per lb., and 75% phosphoric acid costs about 6¢ per lb. (before the rapid inflation of 1974 and 1975) there is considerable economic advantage in using phosphoric acid if sourness alone is the controlling factor.

However, the more expensive fruit acids enhance the special flavor of certain fruit drinks. Also, if economic considerations alone dictated acid choice, the cheapest acid is sulfuric or hydrochloric. Neither is used to any extent, however, because their addition to food is regarded as questionable by some nutritionists. On the other hand, phosphoric acid is cheap, and phosphates are not considered harmful; as shown later they can actually be beneficial to health.

For Nutrient Supplements

As a boy in Southern China, I often accompanied my parents when they brought gifts to friends with newborn babies. To supplement their diet, the new mothers always ate a stew of pigs' feet which had cooked for many hours in vinegar and sugar. This stew was kept cooking and ready all day on the stove, and as visitors we were always offered some, and I remember how delicious it tasted.

My recollection of these visits wasn't stirred until many years later when I suddenly realized that those Chinese mothers were supplementing their diet with calcium and phosphorus—two important minerals in bones, teeth, and milk. The acetic acid in the vinegar converted the calcium phosphate in the bones of the pigs' feet into a soluble and assimilable form. American mothers and pregnant women supplement their diets also—not with stew but with tablets of tasteless calcium phosphates.

Calcium and phosphorus are important not only to humans, but to animals and plants as well. Plants obtain them from the soil, and animals eat the plants. Humans eat both plants and animals and metabolize the calcium and phosphorus to build bones, teeth, muscle tissues, and nerve cells. The non-cellular bone structure of an average adult consists of 60% of some form of tricalcium phosphates; the teeth, 70%. An average person, therefore, carries in his body around seven to eight pounds of tricalcium phosphate. As calcium phosphates are used in various bodily functions, they are replenished by a continuous cycle. Used phosphorus

Table I.   Common Calcium Phosphates

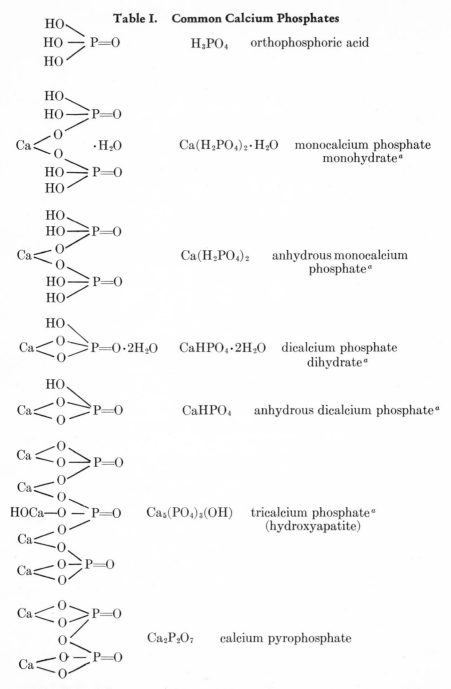

H_3PO_4 orthophosphoric acid

$Ca(H_2PO_4)_2 \cdot H_2O$ monocalcium phosphate monohydrate[a]

$Ca(H_2PO_4)_2$ anhydrous monocalcium phosphate[a]

$CaHPO_4 \cdot 2H_2O$ dicalcium phosphate dihydrate[a]

$CaHPO_4$ anhydrous dicalcium phosphate[a]

$Ca_5(PO_4)_3(OH)$ tricalcium phosphate[a] (hydroxyapatite)

$Ca_2P_2O_7$ calcium pyrophosphate

[a] The prefixes mono, di, and tri used with calcium phosphates indicate that one, two or three of the hydrogens in phosphoric acid are replaced by calcium.

is carried by the blood to the kidneys and excreted in urine, mainly as sodium ammonium phosphate ($NaNH_4HPO_4$).

An average adult eliminates the equivalent of about 3 to 4 grams of phosphoric acid (H_3PO_4) a day into the sewage system. Some of it eventually returns to the soil where plants absorb it and begin the cycle again. However, some excreted phosphorus finds its way into lakes, streams, and waterways through sewage effluents, and causes pollution. This problem is discussed in more detail in Chapter 8.

Nature's cycle of moving calcium and phosphorus from soil to plants and animals and man and back to soil again is too slow to satisfy man. Thus, for human and animal consumption calcium and phosphorus compounds are added as mineral supplements and to stock feeds; for plants, they are added as fertilizers. Before synthetic fertilizers were available, Indians buried a fish in each hill of corn to supply the plants with the calcium and phosphorus from the bones and other parts of the fish; nitrogen was supplied from the degradation of the proteins in the rotting flesh.

Calcium Phosphates for Humans. Calcium phosphates suitable for human consumption in foods are prepared by adding lime to very pure phosphoric acid solution. In this reaction, calcium from the lime (CaO) replaces the hydrogen in phosphoric acid. Since hydrogen ion is monovalent and calcium ion is divalent, each calcium ion can replace two hydrogen ions. For calcium orthophosphates, the particular calcium phosphate compound obtained depends largely on the proportion of lime added to the phosphoric acid. To obtain a compound in its highest purity, one must adjust the ratio of the reacting compounds and adjust reaction conditions to favor formation of that compound. Table I shows the common calcium phosphates, their formulas, and their relationships to orthophosphoric acid.

Calcium Phosphates for Animals. STOCK FOOD GRADE CALCIUM PHOSPHATE. Quality standards for calcium phosphates used in baking acid, mineral supplements, or dentifrices are high. The products must meet all the requirements of state and federal food laws. Because animal life is not valued as highly as human, the specifications for calcium phosphates to be used as mineral supplements for animals are not nearly as strict. Accordingly, the methods for making calcium phosphate for animal consumption are not as stringent.

Stock food grade dicalcium phosphate can be made by hydrating hot lime (CaO) with water and allowing the pasty result ($Ca(OH)_2$) to react with phosphoric acid of 75–80% concentration. The reaction gives off heat, bringing the temperature to about 100°C. Some water evaporates off as steam, and the reaction mixture is dried and ground in a mill. The phosphorus content of this product is almost 21%, close to the theo-

Herefords at feed bunk. Dicalcium phosphate is used as a mineral supplement in animal feeds.

retical value of 22.8% for anhydrous dicalcium phosphate. It consists of monocalcium phosphate, tricalcium phosphate, and unreacted lime, with anhydrous dicalcium phosphate as the main component.

Commercially, feed formulators prefer a product with a phosphorus content of 18.5%. To meet this requirement, enough ground calcium carbonate (limestone) is added to the original product to lower the phosphorus content of the final composition to the desired level.

Another method for making stock feed dicalcium phosphate involves the reaction of hydrated lime with phosphoric acid produced by the wet process. Dicalcium phosphate dihydrate containing 18.5% phosphorus is obtained directly. Since it is prepared from the impure wet process acid, it is contaminated with the usual metal salts present in the acid. These impurities do not seem to be harmful to animals.

In another manufacturing method finely ground limestone reacts directly with phosphoric acid of 67% concentration to form the dicalcium phosphate dihydrate. The use of limestone eliminates the cost of converting the limestone to lime. However, since limestone is not as reactive as lime, the mixture must be stored for about 24 hours to complete the reaction. This product also has a phosphorus content close to 18.5%. In

1972, the United States produced 1.54 billion pounds of 18.5% phosphorus dicalcium phosphate.

Farmers and cattlemen buy the stock food grade dicalcium phosphate and add it along with other mineral supplements and vitamins to the feed for their animals as dietary nutrients.

DEFLUORINATED PHOSPHATE ROCKS. Phosphate rock is a cheap source of calcium and phosphorus. Unfortunately, it cannot be used directly as stock food supplement because most of the naturally occurring phosphate rocks have a composition close to that of the mineral fluorapatite $(Ca_5(PO_4)_3F)$. They contain from 2 to as much as 4.5% fluorine, which is regarded as unhealthy for animals. Also, fluorapatite is inert, and cattle, sheep, and hogs cannot convert its calcium and phosphorus into usable form. One solution to this problem is to extract fluorine from the phosphate rock. The resulting product is defluorinated phosphate rock.

Several commercial methods are used to defluorinate phosphate rock. Most involve the heating of finely ground phosphate rock to 1200°–1400°C. Heating is done in the presence of water vapor with added phosphoric acid, sodium compounds, and silica. The fluoride is evolved as the volatile silicon tetrafluoride (SiF_4) and hydrogen fluoride (HF). By general agreement among producers the finished product must have less than one part in 10,000 of fluoride for every percent of phosphorus in the product. The phosphorus content of the defluorinated rock is adjusted to 18.5% by adding phosphoric acid during the defluorination process.

Various methods can be used to measure the value of the calcium and phosphorus to the animals. The American Association of Agricultural Chemists specifies determination of the solubility of the phosphate (calculated as percent of the total P_2O_5) which is dissolved from a 1-gram sample in 100 cc of neutral ammonium citrate solution. It is assumed that what is soluble in an ammonium citrate solution should also be soluble in the digestive system of the animal.

The most reliable method, however, is the so-called bioassay. Chickens are fed food that is fortified with the phosphate in question. After a specified time, the bones of these chickens are analyzed for total calcium and phosphorus content. If the calcium phosphate added is in an available form, it shows positively in higher than normal bone content.

The test must be carried out under carefully controlled conditions with dozens of chickens in each test group. All the chickens must be fed "off-litter," *i.e.*, they are not allowed to eat their droppings. (Chickens have poor digestive systems and eat their droppings over and over again to get the nutritive value from their feed.)

PHOSPHORIC ACID. Phosphoric acid is also used as a nutrient supplement in animals feeds in combination with molasses. In addition to supplying added phosphorus nutrient value, it also speeds up the molasses, that proverbial slow poke. "Slow as molasses in January" is not an exaggeration. I have seen crude molasses that is slow even in July, but when phosphoric acid is added, it reduces the viscosity and stickiness, thus making both storage and handling easier. This mixture makes a very good animal feed supplement.

Phosphate Leavening Acids

History of Chemical Leavening. The use of leavening agents like yeast to make sponge-like baked goods dates back to the early Egyptians. During its fermentation reaction with the simple sugars in the dough, yeast generates alcohol and carbon dioxide gas (CO_2). These gas bubbles permeate the soft, pliable dough and make it swell—a process that usually takes several hours. During the exodus from Egypt, the Hebrews did not have time to wait for their bread dough to rise; they only had time to bake unleavened dough; the result was matzos.

Yeast is still important in today's baking industry. Most bread, rolls, and coffee cakes are leavened with it. However, the use of chemical agents for leavening dough has grown to such an extent that by 1968–1969 the volume had reached one hundred million pounds a year in the United States alone.

Chemical leavening involves the action of an acid on sodium bicarbonate (baking soda, $NaHCO_3$) to release carbon dioxide gas. Baked goods are prepared mostly from wheat flours that contain the protein, gluten. Gluten is fairly tough and rubbery and can be stretched into films that have a high capacity to retain gas. In preparing dough for baking, the kneading or mixing process disperses the gluten in thin films throughout the system. These films later hold the small nuclei of carbon dioxide bubbles generated by the reaction of chemical leavening agents (or by yeast action and by the air incorporated in the mixing process).

When dough is baked, the gas cells expand as the temperature increases. More gas diffuses into these cells as further chemical reaction takes place in the leavening system, causing the dough to rise further. Finally, at around 160°–170°F (71°–77°C) the gluten sets, and the starch in the flour gels. The baked goods expand very little more during the remaining 10 minutes or so in the oven.

The small holes found in breads, cakes, pancakes, or biscuits are made by carbon dioxide gas. The walls of those holes are cooked gluten films. The spaces between those holes contain gelled starch and other baking ingredients. These holes make the baked goods light and tender while the taste and other properties that characterize cakes or biscuits are the result of the other ingredients.

During mixing and in the early stage of baking, the gas-retaining gluten cell walls are relatively weak. They do not become rigid until set by heat near the end of the baking process. Thus, during the first few minutes of baking, a cake can collapse from excess vibration. It is no wonder that housewives do not like to have their children playing in the kitchen when they have a cake baking in the oven.

The first synthetic chemical system for leavening was introduced in the 1850s. It used cream of tartar (potassium acid tartrate, $KHC_4H_4O_6$) obtained from leftover sediment in the manufacture of wine as the leavening acid to be used with sodium bicarbonate. Soon, monocalcium phosphate was introduced as a leavening acid. The new composition of monocalcium phosphate–sodium bicarbonate baking powder was invented by a Harvard professor of chemistry, E. N. Horsford. His monocalcium phosphate was crude since his process involved the reaction of partially charred bones with sulfuric acid to form phosphoric acid with calcium sulfate ($CaSO_4$) as the by-product. To concentrate the material, Horsford boiled the liquid phosphoric acid which he had separated from the solid calcium sulfate by filtration. A precise quantity of bone ash (crude tricalcium phosphate) was then added to form crude monocalcium phosphate. The resulting moist substance was dried with flour or starch. After further drying for a few weeks, it was ground into dry acid phosphate granules to be used with sodium bicarbonate as baking powder. Horsford's process now seems both ingenious and somewhat quaint.

Monocalcium Phosphate Monohydrate. Refined grade monocalcium phosphate monohydrate, $Ca(H_2PO_4)_2 \cdot H_2O$, is a versatile chemical. Housewives use it in biscuits, adults use it in effervescent headache tablets, and boys use it to shoot off toy rockets.

This phosphate monohydrate is now prepared industrially by adding hot lime (a base) to 75% phosphoric acid in a controlled volume of water. In this order of addition, excess acid is always present since the object is only to neutralize one of the three hydrogens in the phosphoric acid with one equivalent of calcium. The reaction temperature is usually kept between 75° and 110°C and always below 140°C. Above 140°C, anhydrous monocalcium phosphate begins to form.

The reaction for the formation of monocalcium phosphate monohydrate from lime and phosphoric acid in water is:

$$2H_3PO_4 + CaO \xrightarrow{\quad H_2O \quad} Ca(H_2PO_4)_2 \cdot H_2O$$

Monocalcium phosphate prepared this way is about 90% pure; contamination comes mainly from dicalcium phosphate. This impurity is present because during the reaction, a small portion of the monocalcium phosphate formed undergoes a disproportionation reaction to form dicalcium phosphate and phosphoric acid:

$$Ca(H_2PO_4)_2 \cdot H_2O \longrightarrow CaHPO_4 + H_3PO_4 + H_2O$$

The phosphoric acid formed, of course, reacts with lime to form more monocalcium phosphate. This reaction also occurs when monocalcium phosphate monohydrate is dissolved in water. In other words, the compound does not dissolve in water to give a solution of pure monocalcium phosphate. This phenomenon is called incongruent solubility. The presence of up to 10% dicalcium phosphate seems to have no harmful effect on any of the known applications for monocalcium phosphate monohydrate.

Since monocalcium phosphate is actually phosphoric acid that has only one of its hydrogen ions neutralized with one equivalent of calcium, the compound is still an acid although an easy-to-handle, edible, solid acid. Most of its uses are based on this acidic property. The most important use for monocalcium phosphate monohydrate is as a leavening acid in baking.

WHAT IS STRAIGHT BAKING POWDER? Although several kinds of baking powder are on the market today, very few still use cream of tartar as the acid. Some baking powder used today is prepared from monocalcium phosphate monohydrate and sodium bicarbonate mixed in with 37–40% starch (the starch keeps the acid separate from the sodium bicarbonate until it is used in the dough). It is called straight phosphate baking powder.

Because of the fast reactivity of monocalcium phosphate monohydrate, phosphate baking powder liberates two-thirds of the available carbon dioxide gas during dough mixing. Some of the gas stays in the dough, but a large part of it is lost to the atmosphere and is unavailable for leavening. As a consequence, the straight phosphate baking powder has only a minor share of today's baking business.

When monocalcium phosphate monohydrate in straight phosphate baking powder reacts with sodium bicarbonate in a dough, the inorganic chemists like to assume that these compounds react as if they were in a pure water solution; they write the equation:

$$3\ CaH_4(PO_4)_2 \cdot H_2O + 8\ NaHCO_3 \longrightarrow$$
$$8\ CO_2 + Ca_3(PO_4)_2 + 4\ Na_2HPO_4 + 11\ H_2O$$

However, chemists who work with flour know that the reaction is not quite this simple. All they will concede is that carbon dioxide gas is generated because they know that many side reactions also occur between the acid and the baking ingredients and between the acidic impurities in the flour and sodium bicarbonate.

WHAT IS COMBINATION BAKING POWDER? Baking powder prepared from monocalcium phosphate monohydrate in combination with a slow acting acid is called combination baking powder. For household use, the slow acid is generally sodium aluminum sulfate, $NaAl(SO_4)_2$. In 1969 about 85% of all household baking powders were of this type. The sodium aluminum sulfate, commonly called SAS, is believed to react with sodium bicarbonate in this way:

$$NaAl(SO_4)_2 + 3\ NaHCO_3 \xrightarrow{\quad H_2O \quad} 3\ CO_2 + Al(OH)_3 + 2\ Na_2SO_4$$

In a dough system, the reaction does not begin until heat is applied.

A typical combination powder contains 28% sodium bicarbonate, 10.7% monocalcium phosphate monohydrate, 21.4% SAS, and 39.9% starch. When it is used in dough, about one-third of the carbon dioxide (CO_2) is liberated during mixing. It is generated by the action of the monocalcium phosphate monohydrate on some of the sodium bicarbonate and creates the gas nuclei for later expansion. SAS then takes over the leavening action during baking.

However, SAS has its shortcomings; it reacts too slowly, and it continues to react with sodium bicarbonate, liberating carbon dioxide, even after the gluten and starch have gelled and set. When this happens in a biscuit, the side walls split open. SAS can also impart a bitter taste to the baked goods. As manufactured today, SAS contains iron as an impurity; this prevents its use in self-rising flours and prepared mixes because iron, as a catalyst for the rapid oxidation of fats, accelerates the development of rancidity.

SELF-RISING FLOUR AND ALL-PURPOSE FLOUR. Self-rising flour was introduced around 1873 by pre-mixing the baking powder components with flour. Since the original monocalcium phosphate monohydrate varied widely in its composition and quality, this early pre-mix did not find wide acceptance. However, as its manufacture improved, the quality of the self-rising flour prepared from it also improved.

High-quality monocalcium phosphate monohydrate as we know it today was introduced into self-rising flour in the early 1930's. Because this acid is so active, the moisture in the flour, as well as that which seeps into the box from the atmosphere, causes it to react slowly with sodium bicarbonate; thus storage life of self-rising flour is only about three to four months.

Since the product is convenient, many housewives prefer it to regular flour and separate baking powder. At present, the all-purpose self-rising flour contains around 1.375 pounds of sodium bicarbonate, 1.75 pounds of monocalcium phosphate monohydrate, and 2.25 pounds of salt per 100 pounds of flour. Because of the relatively short shelf-life of the product, other better and slower-acting acids are being developed to replace monocalcium phosphate monohydrate. Some of these new replacements are discussed later in this chapter.

Monocalcium phosphate monohydrate is also added in small quantities (0.25–0.75%) to all-purpose flour. This practice began in the 1920's for flour used in the popular sour-milk biscuits. Unfortunately, some batches tasted like soap. Investigation showed that this alkaline taste occurred when the housewife used too much sodium bicarbonate. The acid used for leavening in sour-milk biscuits is lactic acid in sour milk. However, sour milk varies in acid content from batch to batch. A housewife isn't aware of this, however, and she uses the same amount of sodium bicarbonate for the same measured amount of liquid. If the sour milk has a low lactic acid level, soapy tasting biscuits from unreacted bicarbonate are the result.

Chemists found that when monocalcium phosphate monohydrate was added to flour, it compensated for any acid deficiency in the sour milk. Later it was discovered that the added phosphate had other advantages: it modified the gluten so that a softer and more plastic dough was obtained; it prevented "rope development" in the yeast dough. (Rope is caused by the intrusion of bacteria, specifically *Bacillus mesentericus*, which makes the bread stringy or "ropey." Other compounds such as calcium propionate are used widely as preservatives for this purpose.) Monocalcium phosphate also supplies extra calcium and phosphorus, which are valuable nutrient supplements. More than 75% of the all-purpose flour sold today contains added monocalcium phosphate monohydrate. The phosphated flour, therefore, is also present in yeast-leavened baked goods.

BREAD IMPROVER. Monocalcium phosphate monohydrate is also used as the source of calcium and phosphorus for "bread-improver" compositions. A typical composition may contain 7.5% ammonium sulfate, 50.0% monocalcium phosphate, 0.3% potassium bromate, 20.0% salt, and 22% starch. These compositions are used extensively by large bakeries to improve yeast-leavened products—*i.e.*, they stimulate the growth of yeast. About 0.5–0.75% of the composition is added to the flour.

IN EFFERVESCENT TABLETS. Another important use for monocalcium phosphate monohydrate is as the ingredient in effervescent tablets that give off bubbles when added to a glass of water. This bubbling is the result of the quick release of carbon dioxide gas by the action of acidic monocalcium phosphate on sodium bicarbonate—a kind of baking powder in tablet form. This chemical reaction has also provided considerable enjoyment to children. They use the rapidly generated gas to power small rockets. Children with knowledgeable and indulgent fathers sometimes raid their mother's baking powder for this purpose. If they forget to put the lid of the container back on tight, moisture seeps into the baking power. If the powder is later used to bake a cake, the cake will rise to only half its normal height.

Coated Anhydrous Monocalcium Phosphate—the Improved Phosphate Leavening Acid. The anhydrous form of monocalcium phosphate, $Ca(H_2PO_4)_2$, has been known for a long time. However, it was never used as a baking acid because it is hygroscopic; it absorbs moisture on the surface of its crystals and causes them to undergo the disproportionation reaction to form dicalcium phosphate and free phosphoric acid. The free phosphoric acid formed is even more hygroscopic, absorbing more water and causing more decomposition. This decomposition reaction is identical to the incongruent solubility of monocalcium phosphate in water described previously. In a baking powder composition the phosphoric acid formed would also react with the sodium bicarbonate and result in lowered activity.

In the 1930s, almost a century after the origin of chemical leavening, a breakthrough in research on anhydrous monocalcium phosphate occurred. Julian Schlaeger, a chemist, was working on the anhydrous material in the laboratories of Victor Chemical Works, now a part of Stauffer Chemical Co. He and his colleagues developed a delayed-action anhydrous monocalcium phosphate by forming an extremely thin layer of an almost insoluble glassy phosphate on the surface of the fine phosphate particles. The product was given the tradename V-90, and the discovery was hailed by the baking industry as "the most outstanding development in 100 years of use of chemical leavening agents."

Courtesy Stauffer Chemical Co.

Biscuits in left stack were made with anhydrous coated monocalcium phosphate and are higher in volume than those made with regular monocalcium phosphate in the right stack.

I asked Schlaeger, a former colleague, how he came upon this discovery. He hold me that slowing the reactivity of monocalcium phosphate monohydrate had been the major problem for years. In early attempts he tried to coat the tiny particles of the material with a wax or lacquer. Practically speaking, these coatings were useless. If the coating were too thick, the product would not react with the sodium bicarbonate even when mixed with moist dough. Too thin a coating usually resulted in incomplete coverage of the particle, and that was as effective as no coating at all.

This research was done in the 1930s, long before encapsulation—a technique which permits a uniform coating of almost any desired thickness to be applied to a particle. Even if such technology had been available, the cost would have been prohibitive in this case. After all, monocalcium phosphate is relatively inexpensive—7 to 9¢ a pound as of 1971–72. The cost of encapsulation, even today, is many times that.

Schlaeger thought that it might be possible to form a thin coating of calcium acid pyrophosphate or glassy calcium metaphosphate by heating the monocalcium phosphate at a high temperature. Formation of calcium acid pyrophosphate would eliminate one mole of water from two moles of monocalcium phosphate:

monocalcium
phosphate

calcium acid
pyrophosphate

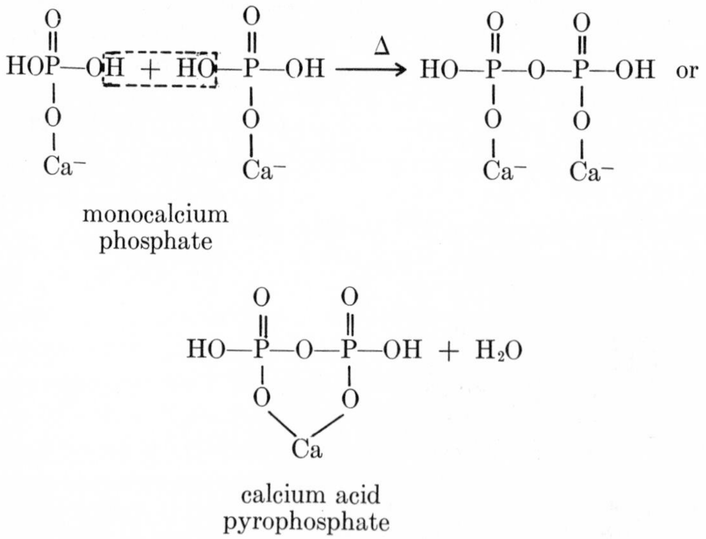

Formation of calcium metaphosphate would involve the elimination of the same number of moles of water as the number of moles of monocalcium phosphate. Such a product should be a high molecular weight polymeric glass. The overall reaction for the formation of this compound from monocalcium phosphate can be shown as:

$$(n+2)\ \ \underset{\underset{\displaystyle Ca^-}{\displaystyle |}}{\underset{\displaystyle O}{\underset{\displaystyle |}{HO-\overset{\overset{\displaystyle O}{\displaystyle ||}}{P}-OH}}} \xrightarrow{\ \Delta\ } HO-\overset{\overset{O}{||}}{\underset{\underset{Ca^-}{|}}{\underset{O}{|}}{P}}-O-\left[\overset{\overset{O}{|}}{\underset{\underset{Ca^-}{|}}{\underset{O}{|}}{P}}-O\right]_n-\overset{\overset{O}{|}}{\underset{\underset{Ca^-}{|}}{\underset{O}{|}}{P}}-OH$$

From general knowledge of this type of compound, a temperature of 200°C would be needed before an appreciable amount of monocalcium phosphate would convert to calcium acid pyrophosphate; for the conversion to calcium metaphosphate, an even higher temperature would be required. Since the desire was to convert only the surface of the particles, this research was exploring new chemical frontiers.

When formed, calcium acid pyrophosphate has two less acidic hydrogens and is less water soluble than monocalcium phosphate; thus, it should have a less tart, acidic taste. The calcium metaphosphate glass should be insoluble in water, and since it doesn't contain acidic hydrogen ions except for one at each end of the long chain, it shouldn't have any tart taste. This simple taste test was used as a guide by Schlaeger to follow his coating experiment. In his exploratory work he heated ordinary monocalcium phosphate monohydrate produced at the Victor plant. Later we will see that the use of this source of monocalcium phosphate was fortuitous.

The temperature chosen for heating was the arbitrary range 210°–220°C; it was based solely on the sixth sense that often guides researchers. Schlaeger knew that when monocalcium phosphate monohydrate is heated to above 140°C, it loses its water of hydration, and the anhydrous material forms. At 220°C, however, he was exploring new chemistry. According to plan, he tasted the product periodically to see if the immediate tart taste had disappeared. After 30 minutes of heating it had, and this was an encouraging sign. This discovery was made at around 4:30 p.m., time to go home. Schlaeger had his assistant determine the neutralizing strength of the heat-treated material. (Neutralizing strength is the measure of the capacity of a baking acid to react with sodium bicarbonate. It is expressed in terms of parts by weight of the acid required to neutralize exactly 100 parts of sodium bicarbonate.) Self-rising flour was then made with this material. As a control, he also made a self-rising flour containing the regular monocalcium phosphate monohydrate. When he took those two formulations home, made them into biscuit dough, and baked them in his wife's oven, a new baking acid was born.

The manufacturing process developed subsequently for the large-scale production of the new baking acid involves the reaction of lime with a strong phosphoric acid at around 140°–175°C. The reaction is done in a batch mixer equipped for efficient mixing. The product is a dry powder of minute crystals of anhydrous monocalcium phosphate. These crystals are then heat treated at approximately 200°–220°C.

Further development seemed fairly straightforward. Large batches were made, and the products were evaluated in different baking formulations. These baking tests gave excellent results. All the biscuits made under carefully controlled conditions showed a volume increase of approximately one-third over those made with ordinary monocalcium phosphate. Similarly, self-rising flour compositions lasted much longer on the shelf without deterioration by absorbed moisture.

One day, without warning, batch after batch of the newly produced material failed; it was no longer superior to ordinary anhydrous monocalcium phosphate. All hands were recruited to investigate this problem. Russell Bell, the microscopist on the project, told me that he spent endless hours peering through his lenses searching for some difference between the minute crystals of the good and bad material. Measurements of the indices of refraction of the crystals showed values all of which characterize only anhydrous monocalcium phosphate; the good and bad crystals were identical by this test. (The fine crystals were all smaller than 200 mesh, meaning particles with diameters less than 0.0029 inch.)

The first indication of some physical difference came finally when Bell watched the crystals dissolve in water under his microscope. The ordinary anhydrous monocalcium phosphate dissolved quite rapidly, leaving hardly any residue. This was also true with the unsatisfactory heat-treated material. However, material from the good heat-treated batches dissolved much more slowly. Even more important, there remained on the microscope slide only very fine transparent empty glassy shells having the shape of the original crystals. These glassy shells were then collected and analyzed. Besides phosphorus and calcium, they contained large amounts of potassium, aluminum, sodium, and a little magnesium along with traces of other minor elements.

Apparently the good heat-treated material was prepared from phosphoric acids containing the four elements as impurities. A check of the phosphoric acids used as raw material showed that that was indeed the case. During that particular period, Victor Chemical Works was making a transition from manufacturing phosphoric acid from phosphorus produced by the blast furnace process to phosphorus from the new electric furnace. The blast furnace acid was far less pure and contained appreciable amounts of the four elements. To check on their idea, the Victor chemists made new batches of monocalcium phosphate monohydrate from the pure electric furnace acid but first added small amounts of potassium, aluminum, sodium, and magnesium. When this was heated, they obtained excellent delayed-action baking acid. Since then some of these trace metal impurities were added to phosphoric acid and became part of the new process.

It was fortunate that for his first experiment Schlaeger had used the monocalcium phosphate monohydrate produced at Victor's own plant and made from the blast furnace acid. If he had used the material from the pure electric furnace acid, V-90 (as the product was subsequently trademarked) might still be undiscovered. Later research showed that the coating is indeed a form of glass with some crystallinity. The major component seems to be a mixed potassium, aluminum, calcium, and magnesium metaphosphate. This composition is formed by the dehydration of the mixed mono-metal acid phosphates. It was later shown that to form calcium metaphosphate as originally envisioned, a much higher temperature would have been needed. Thus the discovery of coated anhydrous monocalcium phosphate did not turn out as originally theorized; it was much better because of unsuspected impurities.

One may speculate that the following sequence of reactions occurs during manufacturing. When lime reacts with phosphoric acid, pure monocalcium phosphate (being less soluble than, for example, monopotassium phosphate, monosodium phosphate, monomagnesium phosphate, and monoaluminum phosphate) crystallizes out. The impurities are concentrated in the mother liquor, the liquid phase. When the whole product is dried at 140°–175°C; the mother liquor containing the four metal acid phosphates dries uniformly on the surface of the pure anhydrous monocalcium phosphate crystals. Heating converts this metal–acid–phosphate coating into a continuous, autogenous, glassy, substantially water-insoluble metaphosphate coating. When the anhydrous monocalcium phosphate is dissolved from the interior of the crystal, the hollow shell coating left behind can be seen under a microscope. Crystals of monocalcium phosphate monohydrate must be precipitated originally in such a small size that after heating they are much smaller than 200 mesh. The particles must be produced in the correct size to begin with for baking purposes. If they are too large, they cannot be ground finer since grinding destroys the coating.

The biggest advantage of the coated anhydrous monocalcium phosphate is its resistance to attack by atmospheric moisture. It is this property that makes possible the formulation of self-rising flour with a long shelf-life. Food processors use it in prepared cake, pancake, waffle, and biscuit mix formulations.

In dough, the coating delays the start of the reaction with sodium bicarbonate, and once gas generation has started, it slows it down. Experiments using V-90 in baking powder to make a biscuit show almost no reaction during the first few minutes of dough mixing. Only about 15% of the carbon dioxide is released during mixing and 35% more during 10–15 minutes of waiting at the bench before baking. As we have seen, monocalcium phosphate monohydrate generates about two-thirds of its gas at the dough-mixing stage alone. The delayed action allows the gluten to become saturated with water and forms films to trap the released gas during the bench action period. This then permits the rest of the leavening gas to be liberated in the early part of the baking. In fact, the coated anhydrous monocalcium phosphate promotes complete leavening action before the gluten and starch set. Each biscuit made with the new phosphates has greater volume, is fluffier, and more appetizing because of a minimum of splits on the side wall.

Commercial introduction of coated anhydrous monocalcium phosphate in 1939 revolutionized the baking industry. A much greater variety of prepared mixes is now available for housewives in the supermarkets, and some say they make better baked goods.

Dicalcium Phosphate as a Leavening Acid. Besides its use in toothpaste, dicalcium phosphate dihydrate (DPD) (*see* Chapter 5) is also used as a leavening acid. This may seem surprising since one reason it is used in toothpaste is its inactivity, insolubility, and indifference to the other compounds in the mix. The immediate pH of a water slurry of dicalcium phosphate dihydrate is around 7.4–7.5 or slightly alkaline. When it is used as a leavening acid, chemists take advantage of one of its undesirable properties which had to be corrected before it could be used in a toothpaste. DPD, when unstabilized, dehydrates and disproportionates in the presence of water into hydroxyapatite and phosphoric acid (or maybe to the acidic monocalcium phosphate). Heat triggers the reaction. It's the acid from this process—whether as free phosphoric acid or as acidic monocalcium phosphate—which acts as a leavening acid.

Naturally, such a slow-reacting, heat-dependent leavening action would be useless for pancakes or waffles, which require quick leavening; but it is very useful in cakes. A cake takes about 30 minutes to bake, and the setting temperature—that is, the temperature at which the cake becomes firm—is high (about 160°–170°F (71°–77°C) inside the cake). For cakes most of the leavening action is accomplished by beating air into the batter. In many recipes sodium bicarbonate is also added. It releases carbon dioxide when heated to 140°F (60°C), leaving an alkaline residue and occasionally some unreacted sodium bicarbonate. Dicalcium phosphate dihydrate releases its acidity at around 150°–160°F (66°–71°C) just before the cake sets. Actually, the released acidity contributes very little to the leavening of the cake. Its main purpose is to neutralize the soapy alkaline taste from the remaining sodium bicarbonate.

Sodium Acid Pyrophosphate as a Leavening Acid. DOUGHNUTS. Sodium acid pyrophophate, SAPP, has an acidic property which makes it useful as a baking acid, and its pyrophosphate ion enables it to sequester (tie up) many metal ions such as iron, magnesium, and calcium. These properties are used to make tender doughnuts and to keep boiled potatoes white.

SAPP is prepared by removing one mole of water from two moles of monosodium phosphate:

$$\underset{\substack{HO}}{\overset{\substack{NaO}}{P}}\!\!\overset{O}{\underset{}{\parallel}}\!\!-\!O\!\overset{}{H}\!+\!H\overset{}{O}\!\!-\!\!\underset{\substack{OH}}{\overset{\substack{O \; ONa}}{P}}\quad\xrightarrow[\text{to }250°\text{C}]{225°}\quad\underset{\substack{HO}}{\overset{\substack{NaO}}{P}}\!\!\overset{O}{\underset{}{\parallel}}\!\!-\!O\!\!-\!\!\underset{\substack{OH}}{\overset{\substack{O \; ONa}}{P}}\quad+\quad H_2O$$

The temperature range of 225°–250°C is critical because as the temperature rises, monosodium phosphate is converted to meta- and polyphosphates at 530°–600°C and 800°–900°C respectively.

SAPP with its two acidic hydrogens is the solid acid in the baking powder used by many commercial bakeries. At room temperature it reacts very slowly with sodium bicarbonate, even after it is mixed in the dough or batter. This slow reactivity at room temperature permits bakers to mix dough or batter in large batches and bake at leisure. The reaction of SAPP with sodium bicarbonate doesn't start until the dough is heated in the oven.

The reason for this slow reaction of SAPP with sodium bicarbonate is not well understood. In water, the two compounds react rapidly, as any water-soluble acid is expected to react with sodium bicarbonate. In a batter, other ingredients apparently interfere. Some believe that this interference is caused by the coating of SAPP particles with insoluble calcium pyrophosphate from the reaction of SAPP with calcium ions from batter ingredients such as milk solids. This speculation is supported partly by the fact that when more SAPP surface area is available (as with finer particles), it is less reactive.

In normal chemical reactions, when no surface coating interference is involved, the finer the size of the reactant particles, the faster the reaction. Also, the reaction between SAPP and sodium bicarbonate is quite rapid in a batter that contains no calcium salts.

The reaction between SAPP and sodium bicarbonate—which won't begin until the batter is heated—is ideal for making doughnuts. The preparation of a cake doughnut (in contrast to a bread doughnut which is raised by a yeast fermentation reaction) is an art that requires a great deal of science. Doughnut batter is not as thick as biscuit dough and is not as thin as cake batter. In doughnut batter it is important that carbon dioxide liberation be controlled during mixing. If too many gas nuclei form during quick deep-frying, they cause the doughnut to overexpand into a highly porous mass that absorbs too much frying oil, resulting in a greasy doughnut.

What is needed is an acid that reacts with the sodium bicarbonate in the doughnut batter only when heated. Such a temperature-triggered reaction should give a sustained release of carbon dioxide gas for one and a half to two minutes. The resulting doughnut is just porous enough to absorb the right amount of fat for good flavor and to have a firm, pleasant looking shape.

In our laboratories at Stauffer Chemical Co. we have made thousands of doughnuts using an automatic machine, mounted on a sensitive scale, that make 30 to 40 dozen doughnuts per hour. We test doughnut formulations containing different baking acids to determine their influence on the absorption of fat by the doughnuts. Since the machine is mounted on a scale, the amount of fat absorbed by a dozen doughnuts can be determined simply by weighing the loss of fat after each dozen doughnuts is fried.

Still, a good doughnut is not only a function of the baking acid and the other ingredients; it also depends on frying conditions. Although our doughnut-making machine is a commercial unit used in specialty shops, it does not quite duplicate the conditions and results of the large

Courtesy Stauffer Chemical Co.

Whether he knows it or not, this smiling chef relies on phosphates to produce the delectable assortment of food shown here. Reading from the left: phosphoric acid is used in the soda drinks; monocalcium phosphate, 1-3-8-, 2-3-8- sodium aluminum phosphate is used in making the biscuits; phosphoric acid gives clarity to the gelatin dessert; sodium acid pyrophosphate in the water prevents after-cooking darkening of sweet and white potatoes; disodium phosphate or sodium aluminum phosphate (Kasal) emulsifies the cheese; the skin of peas and corn will become tough when they are cooked unless polyphosphate is added to the water to chelate the calcium ions present.

machines in doughnut factories which makes tens of thousands of doughnuts an hour. The smallest machine we could find to duplicate the results of our customers' big machines was one that makes 100 dozen doughnuts an hour. For each formulation we test, we make several hundred doughnuts, and one of our problems is trying to get rid of all the doughnuts we make. We can only eat so many.

REFRIGERATED BISCUITS. Refrigerated biscuit dough involves ingenious chemistry. The dough prepared from the flour mix containing SAPP and sodium bicarbonate is rolled into a sheet and then cut into biscuits. This is done at 55°–57°F (13°–14°C) (dough temperature) to slow down the already slow room-temperature reactivity of the system. Cut biscuits are stacked into a fiber (cardboard) can lined inside with a spiral of aluminum sheet in which one section overlaps the next. The ends of the can are then sealed with metal caps.

These canned biscuits are placed in a warm area at between 80°–100°F (27°–38°C) for about one-half hour. The reaction between the SAPP and sodium bicarbonate begins, and the dough swells. This swelling squeezes residual air from the can and presses against the overlapping aluminum foil, sealing the can tight. The can is thus filled with swelled dough under a carbon dioxide atmosphere.

Since the can is under constant pressure, air cannot seep in, and aerobic bacteria fermentation (which depends on oxygen from air) is prevented. Canned biscuits kept at 35°–40°F (2°–4°C)—the normal temperature range of a refrigerator— will keep for two to three months, and the cans will withstand pressures up to 90 pounds per square inch (six times atmospheric). A can which has popped open in the refrigerator of a grocery store has not been stored properly.

COMMERCIAL BAKING POWDER. Commercial bakeries use a baking powder of monocalcium phosphate monohydrate mixed with a slow acting acid. The acid is usually sodium acid pyrophosphate (SAPP) or calcium lactate. The reaction of calcium lactate with sodium bicarbonate is assumed to occur this way:

$$Ca(C_3H_5O_3)_2 + 2NaHCO_3 \longrightarrow CO_2 + 2NaC_3H_5O_3 + CaCO_3 + H_2O$$

The $CaCO_3$ thus formed may react further with monocalcium phosphate monohydrate to liberate more carbon dioxide gas:

$$Ca(H_2PO_4)_2 \cdot H_2O + 2CaCO_3 \longrightarrow 2CO_2 + Ca_3(PO_4)_2 + 3H_2O$$

After the CO_2 gas nuclei are formed, further leavening occurs only when heat is applied. Large batches of dough can thus be prepared at one time and baked at leisure.

The Sodium Aluminum Phosphates—A New Family of Baking Acids.
Sodium aluminum phosphates are such a new family of compounds that

they are not described in today's inorganic chemistry text books. At present the two industrially important members have the formulas $NaH_{14}Al_3(PO_4)_8 \cdot 4H_2O$ and $Na_3H_{15}Al_2(PO_4)_8$. They are crystalline compounds with individual x-ray diffraction patterns. They are excellent leavening acids and in many respects are quite different from the leavening acids discussed earlier.

In the section on calcium phosphates, I discussed the use of mono-calcium phosphate monohydrate as a leavening acid for baking. I also described the invention of coated anhydrous monocalcium phosphate which, through its microscopic crystalline coating, ensures a delayed reaction with sodium bicarbonate. This delayed reaction made possible the formulation of self-rising flour and ready-mixed baking formulations with much longer storage times than previously possible. It also resulted in improved baked goods. In the section on sodium acid pyrophosphate I discussed the importance of SAPP in many baking systems and noted that when it is used in baking, it provides most of the leavening action only after the batter or dough is heated. This property is important for institutions where a large batch, for example, of pancake or waffle batter is prepared in the morning but must remain useful for the rest of the day.

The above summary would imply that all problems concerning leavening acids were solved, but this is far from true. The competitive search for an improved leavening acid continued and it led to the discovery of sodium aluminum phosphates. The first one introduced commercially in the late 1950s is $NaH_{14}Al_3(PO_4)_8 \cdot 4H_2O$. It is referred to as 1-3-8 SALP with the numbers indicating the ratio of the sodium (Na), aluminum (Al), and phosphorus (P) atoms in the molecule. $Na_3H_{15}Al_2(PO_4)_8$, which was introduced later, is called 3-2-8 SALP, with the numbers again indicating the ratio of the three principal atoms. Both compounds can be made by adding the correct ratio of sodium (as sodium carbonate or sodium hydroxide) and alumina (Al_2O_3) to excess, hot concentrated phosphoric acid. On cooling, crystals of the respective compounds precipitate out.

Unlike the monocalcium phosphates and SAPP, the above two compounds are quite complex. As yet we do not know exactly how the H, Na, Al, P, and O atoms are connected—i.e., we have not yet determined their molecular structures. We do, however, know a good deal about their physical and chemical properties and how to apply them.

Since I was associated with the inventors of 1-3-8 SALP, I followed their trials and tribulations. I saw the preparation of the compound in a little beaker, its evaluation in baking biscuits to the larger-scale preparation in a pilot plant, and the complete evaluation in many different commercial baking formulations. All this led finally to the successful commercial production of the compound in a multimillion pound industrial chemical manufacturing plant.

In the development of 1-3-8 SALP many of the original obstacles were technical. For example, the pure compound is quite hygroscopic, absorbing moisture from the air and becoming sticky. One can envision bags of it shipped as a free-flowing powder arriving at the customer's plant as sticky lumps. Also, one would expect that a leavening acid wetted with absorbed water would be much more reactive with sodium bicarbonate. Such reactivity would make a self-rising flour inactive almost instantly. Laboratory experiments, however, showed that when 1-3-8-SALP is incorporated in a self-rising flour, the slight moisture it picks up is beneficial. The fine flour particles stick to it and protect it from reacting with sodium bicarbonate. The problem of its becoming sticky on storage and shipment, however, had to be solved before it could be marketed. Basic research in inorganic chemistry paved the way. The final answer involved treating the compound with a small amount of a soluble potassium salt. The potassium ions displace some of the hydrogen ions from the surface of the 1-3-8 SALP crystal. This changes the surface characteristics of the crystal, reducing its hygroscopicity to a satisfactory level without affecting the compound's function as a leavening acid. The potassium salt treatment along with some formulation modifications have resulted in a product that remains free-flowing even after prolonged storage.

The discovery of the sodium aluminum phosphate leavening acids coincides with the development of newer emulsifier systems (shortenings) for baking. Gas bubble nuclei are formed in the dough or batter by the mechanical mixing-in of air and by the action of the leavening system which generates some CO_2 gas at that initial stage. Much of the gas usually escapes during mixing. In baking, the gas bubble nuclei, which formed initially in the dough, are expanded by heat and also by the diffusing-in of the newly formed CO_2 gas. Thus it was desirable to have a double-acting baking powder. One action generates the gas bubble nuclei during mixing and bench action, and the other generates gas during baking.

With the advent of the new emulsifiers or shortenings, which are used in the "prepared baking mixes," any bubbles which form during mixing are stabilized by these shortenings, and few escape. Thus the need for a gas-generating system that does its work during mixing is much less. What is needed are gas-generating systems that become active during baking to expand the size of the bubble nuclei already formed and that are stabilized by the new shortening—in other words, a delayed-reaction leavening system. Sodium aluminum phosphate acids provide this delayed reaction.

The reaction of 1-3-8 SALP with sodium bicarbonate in a water system is shown below. This equation is based only on the reactants used and the amount of CO_2 gas evolved. The structure of the new sodium aluminum phosphate which is formed has not been completely confirmed.

$$2\ NaAl_3H_{14}(PO_4)_8 \cdot 4\ H_2O + 23\ NaHCO_3 \longrightarrow$$
$$Na_5Al_6(PO_4)_6(OH)_5 \cdot 12\ H_2O + 10\ Na_2HPO_4 + 14\ H_2O + 23\ CO_2$$

When 1-3-8 SALP is used as a leavening acid, it releases its acidity to react with sodium bicarbonate very slowly. Only 20–30% of the totally available CO_2 gas is released during mixing and the bench stage. The remaining 70–80% is released slowly during baking. Experiments have shown that it has a more desirable gas-releasing rate for cakes than sodium acid pyrophosphate.

Sodium aluminum phosphate is the most stable leavening acid available today. Its low reactivity with sodium bicarbonate at room temperature in a batter makes it possible to prepare pancake and waffle batters which can be delivered in wax cartons by dairies. When refrigerated, these batters contain sufficient leavening action to make very good pancakes and waffles even after many days of storage. In fact, such batter is usually destroyed by bacterial action before the leavening action is gone. Biscuit doughs made with self-rising flour containing sodium aluminum phosphate lose very little of their leavening action after 24 hours in a refrigerator.

The above examples show the characteristics of sodium aluminum phosphate as a leavening acid and its superiority over previously known agents. However, SALP is not so perfect as to replace all of the previously known acids. Rather, it supplements them, extending and improving their usefulness. It has made possible leavening systems heretofore unavailable. In fact, much of the sodium aluminum phosphates are used today in combination with such leavening acids as the coated anhydrous monocalcium phosphate. They are used in household and commercial baking powders, in self-rising flours, and in prepared biscuit and cake mixes.

From a baker's point of view, sodium aluminum phosphate, when used as a component in the new leavening acid system, produces baked goods with a finer and more even grain and with a tender and resilient crumb. In flavor-sensitive systems, its blandness allows only the desired flavor to predominate. (SAPP when used as the leavening acid, usually imparts an astringent after-taste known as the pyro taste.)

Phosphates for Dairy Products

Cheese Emulsifiers. DISODIUM PHOSPHATE DIHYDRATE. Disodium phosphate, as its name implies, is phosphoric acid whose two hydrogen atoms have been replaced by sodium atoms. It is prepared by the reaction of one mole of sodium carbonate with one mole of phosphoric acid. Commercial products are the anhydrous form (Na_2HPO_4) and the dihydrate form ($Na_2HPO_4 \cdot 2H_2O$):

$$H_3PO_4 + Na_2CO_3 \longrightarrow Na_2HPO_4 + H_2O + CO_2$$

Of course, the reaction is not as simple as adding the correct amount of sodium carbonate to a water solution of phosphoric acid. In practice, the sodium carbonate used is not basic enough to drive the reaction to completion. It is necessary to supply about 15% of the sodium in the disodium phosphate from the more basic sodium hydroxide. Further, reactant concentration and reaction temperature must be controlled carefully. If too concentrated a solution of reactants is used with too high a temperature, tetrasodium pyrophosphate tends to form as a co-product. As much as 0.5% of tetrasodium pyrophosphate in the disodium phosphate is detrimental to its use in processed cheese, for example.

The largest use for disodium phosphate is as an emulsifier for pasteurized American processed cheese—*i.e.*, to distribute the butterfat uni-

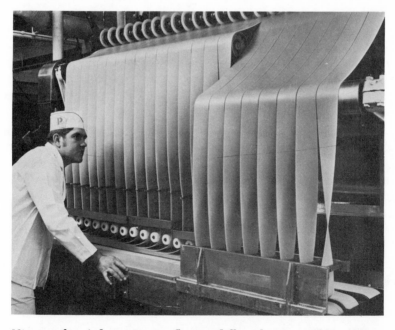

Ninety miles of cheese comes off giant chilling drums in golden ribbons each day at Swift & Co.'s plant in Green Bay, Wis. enroute to automatic slicing and packaging operation. This much processed cheese each day is enough to make nearly 3 million cheeseburgers, or a billion a year. Disodium phosphate finds its largest use as an emulsifier for pasteurized American processed cheese.

formly throughout the cheese. All cheese is the result of controlled fermentation of a precipitated milk curd (milk protein) which contains entrapped butterfat. Blue cheeses, such as Roquefort and Danish blue, are the result of curd fermentation by a specific mold (*Penicillium roquefort* and *Penicillium glaucum*). Cheddar cheese, the most popular in the United States, is produced when milk curd is ripened with a special culture of bacteria. When such spoilage goes unchecked and the bacterial decay continues, aged Cheddar results. However, not all precipitated milk curds, even when inoculated with the same bacteria strain, result in cheese with the same degree of aging; several grades, all with varying flavors, are possible.

To ensure uniform flavor and consistency, the pasteurized cheese process was developed. In pasteurization, various grades of Cheddar of different properties—including some called stinkers—are blended and cooked with added steam at around 160°F (71°C) into a homogeneous mass. Cooking and pasteurization inactivate the bacteria and prevent further degradation or flavor changes. The melted cheese is poured into a container where it solidifies into a loaf, or it is extruded as a ribbon and cut into slices.

In this cooking and pasteurization, an emulsifier is necessary to distribute the butterfat (originally trapped in the milk curd) homogeneously throughout the cheese; otherwise it will separate during cooking or storage. Although sodium citrate is used, the most popular emulsifier is disodium phosphate ($Na_2HPO_4 \cdot 2H_2O$) (DSP) because of its ready solubility in the melted cheese mixture.

The mechanism of the emulsifying action of disodium phosphate (DSP) is not yet completely understood. We do know that it prevents fat globules from separating when cheese is melted. The U.S. Food and Drug Administration permits the use of a maximum of 3% added emulsifier, whether citrate or phosphate, or a combination of both. The practical limit for disodium phosphate use is about 1.8%. Above this level, it tends to crystallize out in the cheese as the dodecahydrate ($Na_2HPO_4 \cdot 12 H_2O$) which has the appearance of glass splinters. Even though it is not harmful, it is difficult to convince a housewife that she is not looking at glass splinters in the cheese.

Raw cheeses used for pasteurized process cheese contain about 30–38% moisture. Since federal regulations allow a maximum of 40% water in processed cheese, this level is attained by adding steam during processing. The emulsifier is used, therefore, to prevent the separation of not only fat but also water.

Pasteurized processed cheese can be made without added emulsifier and still have no separation problems. However, the raw cheese must have the correct natural balance of magnesium citrate and calcium phosphate. Because cows graze on different grasses at different seasons, they produce milk of varying calcium, magnesium, citrate and phosphate content. Phosphates and other ingredients are added to overcome nature's vagaries.

Disodium phosphate as an emulsifier produces a cheese that melts to a gooey mess when heated. However, if pyrophosphate is inadvertently introduced as an impurity, the cheese won't melt at all when it's heated. Unfortunately we don't yet have a good scientific answer for this behavior.

As a cheese emulsifier DSP is also used in combination with insoluble sodium metaphosphate (IMP). Commercial mixtures contain about 40–70% of IMP and 30–60% of DSP (sometimes trisodium phosphate is used in place of DSP). A total of 3% of such a mixture may be used in processed cheese. Since IMP is practically insoluble in water, there has been speculation that its presence in the mixture acts only as a diluent. Test results, however, have confirmed that IMP does contribute to some degree as a cheese emulsifier. Tradenames such as Emulsifos and Kasomet are used commercially for IMP-sodium phosphate cheese emulsifier compositions.

KASAL—A SODIUM ALUMINUM PHOSPHATE. The third commercially important sodium aluminum phosphate, although it has the empirical formula $Na_{15}Al_3(PO_4)_8$, is not a pure crystalline compound. It can be made by the reaction of sodium hydroxide, alumina, and phosphoric acid in the ratio indicated by the formula. Drying the reaction mixture yields a free-flowing white powder. Since it contains no acidic hydrogen, it is not a leavening acid but an efficient emulsifier for processed cheese. Its trade name, Kasal, is derived from the German word Käse for cheese.

Robert Lauck, one of the inventors of Kasal, told me how he happened to discover this compound. In connection with another problem, he was studying the dispersing action of the milk protein, casein, in water. Since he was aware of the action of phosphates on proteins, he had tried various alkaline phosphates as dispersing agents and wanted to try an aluminum phosphate. The 1-3-8 SALP which was available was an acid and not suitable since he needed an alkaline sodium aluminum phosphate. Thus, he neutralized acidic 1-3-8 SALP with sodium carbonate. The resulting product was indeed an effective casein dispersant. He reasoned that since this alkaline sodium aluminum phosphate has some effect on casein, it might also emulsify pasteurized processed cheese.

The important word in the preceding sentence is "might." Chemists use it for their hypotheses all the time, and the probability of converting the term "it might" to "it will" is not very high. In Lauck's case, success was not immediate. After partial successes and partial failures, and with the assistance of a large industrial organization, he finally achieved his goal.

Unlike disodium phosphate dihydrate which is seldom used more than 1.8% because of the possibility of the glass splinter-like crystal formation of disodium phosphate dodecahydrate, Kasal can be used to a maximum level of 3% as permitted by the Food and Drug Administration. Its blandness also permits the cheese flavor to dominate, and thus processed cheese of improved properties and flavor is obtained.

Phosphates in Processed Milk

Disodium Phosphate in Evaporated Milk. Evaporated milk is made by heating raw milk to *ca.* 240°F (116°C) to concentrate it. If it is stored, the milk gels slowly into a mushy solid. If about 0.1% anhydrous disodium phosphate is added before heating, gelation is prevented and the milk remains liquid. Here again phosphate addition helps to maintain the correct balance of calcium and phosphate. It is needed especially in milk produced in the springtime when phosphate content is normally low. Although there are many hypotheses as to why a correct calcium–phosphate balance prevents gelation, a definitive answer is still needed.

Tetrasodium Pyrophosphate (TSPP) in Buttermilk, Chocolate Milk, and Puddings. The ability of TSPP (Chapter 7) to disperse solids accounts for its use in buttermilk. When whole milk is churned to coagulate the cream or butterfat into butter, the residual liquor after fermentation is buttermilk. Lactic acid formed during fermentation is responsible for its sour taste. If the buttermilk is to taste good and look appealing, it must have the right consistency. If the curds agglomerate too much and the buttermilk becomes too thick, TSPP disperses them.

When tetrasodium pyrophosphate is ionized in water, it produces four positively charged sodium ions called cations. It also produces one pyrophosphate ion with four negative charges called an anion:

$$Na_4P_2O_7 \xrightarrow{\quad H_2O \quad} 4\ Na^+ + P_2O_7^{4-}$$
$$\text{pyrophosphate}$$
$$\text{anion}$$

The ability to disperse solids is common to anions having multiple negative charges, such as the pyrophosphate anion. One theory to explain this phenomenon proposes that the solid particles absorb the negatively charged pyrophosphate anions on their surfaces. That is, the pyrophosphate anions, each with the four negative charges, adhere to the solid particles and in effect give all the particles a negative charge. Since negatively charged particles repel each other, the particles disperse.

Thick chocolate milk also uses TSPP. The thickening of milk products is just the reverse of the dispersing action described for buttermilk. In chocolate milk, calcium reacts with the pyrophosphate to form a weak calcium pyrophosphate gel. This gel interacts with milk protein to give it more body. Since milk is often drunk through a straw, the thickness must be limited by adding only about 0.1% tetrasodium pyrophosphate. Unexpectedly, chocolate milk thickened with TSPP also has a richer color and a superior flavor. However, no good theory has been developed to account for these advantages.

Since TSPP reacts with calcium ions and protein in milk to form a gel, it is also used in instant puddings that don't require cooking. These formulations usually contain precooked starch, sugar, salt, flavoring, TSPP, and some added soluble calcium salt such as calcium acetate to supply extra calcium ions. The extra calcium reacts with TSPP to produce more calcium pyrophosphate gel which then reacts with the protein in milk to give a firm gel (pudding). (This contrasts with chocolate milk where only thickening is desired, in which case the calcium in the milk itself is sufficient.)

Phosphates in Meat Products

Sodium Phosphates in Ham Curing. Both disodium phosphate and sodium tripolyphosphate are used to cure hams. The flavor of ham, besides that imparted by smoking, is the result of controlled bacterial decay. The bacteria comes from the environment, and the taste of the decayed meat is what we recognize and enjoy as ham flavor.

Normally ham is fermented or cured in a pickling (salt) solution at around pH 6. I was told that disodium phosphate was originally added to increase the pH to about 6.5–6.8, thus favoring bacterial growth, just enough to improve the ham flavor. About 5% disodium phosphate was added to a brine solution containing about 15–20% salt and other pickling ingredients. This solution was injected into the ham through the pig's arteries until the equivalent of 10% of the weight of the ham was added. The ham was cured with the pickling solution for several days and then smoked.

Unexpectedly, hams treated with this solution were tenderer and juicier than those cured without added phosphate. In normal curing, ham usually loses about 10% of its weight. If water is added, the ham becomes soggy. Ham cured with added phosphate loses almost no weight and is juicy but not soggy.

A few years ago, a consumer group was agitating for the discontinuance of the phosphate pickle process. They complained that producers were selling added water to the public as part of a ham. However, be-

Hams and sausages are pickled with the use of disodium phosphate and sodium polyphosphates. These compounds improve the water retention of the meat and thus increase juiciness and improve the flavor and eating quality.

cause of the quality improvement in ham when water is retained by the added phosphate, the producers won their case.

The favorable results obtained with sodium tripolyphosphate on ham, I was told, were also not expected. Originally sodium tripolyphosphate was added to kill bacteria in sausages and ham. Cured meat products, exposed to air, often develop an unappetizing greenish color; this is the result of split myoglobin, the pigment responsible for the red color of the meat. Myoglobin is split by hydrogen peroxide formed by bacteria that grow on meat. This can be demonstrated by cutting a slice of bologna and pouring hydrogen peroxide on it. A greenish color develops almost immediately. Although sodium tripolyphosphate was added originally to kill the bacteria that generate hydrogen peroxide, it not only improved the color of sausage, but also retained the moisture and fat which impart a desired plumpness and texture to the sausage. Added to ham, it gave a juicier, tenderer product with improved flavor and nutritive properties. When I mentioned to a Jewish friend that sodium tripolyphosphate, the same chemical used in laundry detergent (Chapter 7), is used to pickle ham, she remarked facetiously that now at least she knows ham is clean.

If we compare the original concept for using disodium phosphate to cure ham with that of using sodium tripolyphosphate, it's interesting that in one case the intention was to grow bacteria to improve the flavor while in the other it was to kill bacteria to improve the color. Both approaches are successful in unexpected ways. As a mater of fact, TSPP and tetrapotassium pyrophosphate are also effective for pickling ham. Commercial operations sometimes use a mixture of phosphates. The present hypothesis is that added phosphates somehow increase the water-binding capacity of proteins. The elucidation of the precise mechanism for this phenomenon, however, requires more detailed scientific study.

Potassium Polymetaphosphate in Sausage. Potassium polymetaphosphates are quite different from the sodium polymetaphosphates. The commercial water softener Calgon is a water-soluble sodium polymetaphosphate with 15 to 20 moles of phosphorus for each molecule. There is no counterpart to Calgon in the potassium metaphosphates. The potassium polymetaphosphates are generally quite insoluble in water. They are made by heating monopotassium phosphate to 350°–400°C. If a very high molecular weight potassium polymetaphosphate is desired— average molecular weight of over a million— a temperature of about 500°C is needed.

Physically, potassium polymetaphosphate looks like white asbestos with a fiber-like appearance. It can be soaked in distilled water for a long time at room temperature without turning gummy. In water that contains metal ions it behaves quite differently. For example, it dissolves to give a very viscous solution in water when a little sodium chloride is added.

Solid potassium polymetaphosphate can be visualized as a series of long chains compacted together tightly in a crystalline lattice. This lattice is so tightly packed that the water molecules cannot get in between the long chains. However, when a soluble sodium salt (or other monovalent ion such as lithium) is added, the sodium ions replace some of the potassium ions. Since the sodium atom is not the same size as the potassium atom, the neat, tight symmetry of the chain is disturbed. Now water molecules can enter and pry the chains loose from their crystalline lattice. After leaving the lattice, the chains disperse in water like a gel, which resembles a solution of natural gum, or starch. If, at this point, a bivalent ion such as calcium is added, each calcium ion can replace two of the sodium or potassium ions from two separate chains. The net result is that the chains are tied together with calcium ions; this is called crosslinking.

Such crosslinked material is no longer soluble in water. Since it is a network of many long chains with small holes in between, the holes can hold some water. The finished product is a rubbery gum which can be stretched into a film or squeezed into a round bouncing ball. Such properties are unusual indeed for a completely inorganic compound. Unfortunately, no way has yet been developed to hold the water permanently. It evaporates with time and the product becomes a solid powder.

Potassium metaphosphate, dispersed in water as a clear and viscous gel, is used in sausage manufacture. The meat mixture contains at least 30% fat, sometimes as much as 60%, and added water. The best ingredient for binding, absorbing, and holding the fat and moisture together during smoking or cooking is hot bull meat. A more correct term would be warm bull meat, that is, meat from a slaughtered bull. It is added to the mixture before the meat cells become cold (before *rigor mortis* sets in). The exact mechanism for the action of hot bull meat in sausage is not clearly understood. Meat cells undergo a physicochemical change on cooling. Apparently the cell wall hardens and prevents the full release of proteins, albumen, and hemoglobin as well as naturally occurring organic phosphates and other proteins. When bull meat is dumped into the cutter in the sausage manufacturing machinery, the almost living cells somehow help the hemoglobin, albumen, natural phosphate, and other proteins in the pork to act as a binder for fat and moisture. When the resulting sausage is boiled or cooked it remains juicy, and the fat stays evenly distributed.

However, bull meat is becoming scarce. Many potential bulls are slaughtered while young and sold for veal. To replace it, new formulations containing potassium polymetaphosphate along with tetrapotassium phosphate and tetrasodium pyrophosphates are being patented. However, potassium polymetaphosphate was not yet approved for food use by the Food and Drug Administration in the United States at the time this book was written. It is used in sausages, however, in many European countries.

Phosphates in Seafood Products

Phosphates also improve the quality of freshly processed fish as well as canned seafoods. For example, when a fish is fileted, it usually exudes a slimy liquid, more commonly known as drip, which is a solution of soluble protein. The amount of exudate is increased when the frozen filets are thawed. Prior to freezing, if the filets are dipped in a solution of about 12.5% sodium tripolyphosphate and 4% salt, the drip loss—and thus protein loss—is significantly decreased. This development has led to improved quality in frozen fish filets as well as improved nutritive value.

Many other seafood products such as lobsters, shrimp, crab meat, haddock, cod, and salmon are preserved by canning. Over a period of storage time, crystals of magnesium ammonium phosphate ($MgNH_4PO_4$), known as struvite, form. While they are harmless both physically and

Courtesy Stauffer Chemical Co.

Sodium acid pyrophosphate or sodium polyphosphate used in ¼ to 1½% concentration based on the total moisture content of the canned seafood is sufficient to prevent the formation of struvite crystals (magnesium ammonium phosphate, MgNH₄PO₄) which look like sharp pieces of glass.

nutritionally, they do look like sharp pieces of glass and have resulted in consumer rejection. Addition of 0.25–1.5% of sodium acid pyrophosphate or sodium polyphosphate based on the total moisture content of the canned seafood prevents struvite formation. The effectiveness of these polyphosphates is based on their ability to sequester magnesium so that it is no longer available to form $MgNH_4PO_4$. One advantage of these phosphates is that they are generally recognized as safe in all food products.

Phosphates in Cereal and Potato Products

Cold Water Gel Starch. DISODIUM PHOSPHATE. Disodium phosphate alone or combined with monosodium phosphate is used to make starch for instant puddings and pie fillings. The starch is slurried with a solution of mono- and disodium phosphate at pH 6.1–6.5. After filtration, the starch is heated in a vacuum oven at *ca.* 60°C for a few hours. This treated starch will now form a gel when cold water is added; normally starch only gels on cooking. The cold-water gel from phosphate-modified starch does not become thin on aging or cooking and can be used in desserts as well as industrial sizing for textiles and papers (filling in the pores in the surface). Even though we find phosphate-modified starch useful, we still don't really know how the starch molecule is modified. The phosphate content in this starch is very low—a fraction of 1% by weight of the starch.

CYCLIC SODIUM TRIMETAPHOSPHATE. Cyclic sodium metaphosphate is an interesting compound, but few industrial applications exist for it. One is for treating corn starch to make a product that forms a stable jelly with cold water. The property of such starch is not unlike that produced by disodium phosphate treatment discussed earlier. This starch is used to thicken food products—soups and canned vegetables and fruits—where a relatively clear thickening agent is preferred. It also can be used to formulate pudding mixtures which require no cooking.

Cyclic sodium trimetaphosphate is prepared by heating monsodium phosphate to about 530°–600°C. Three moles of monosodium phosphate join head to tail and lose three moles of water to form a ring.

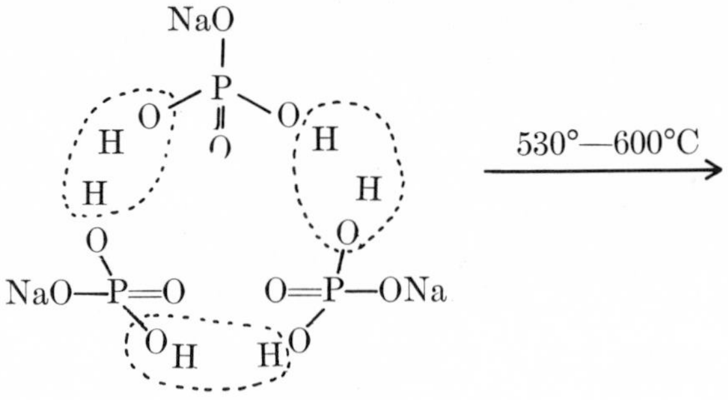

monosodium phosphate

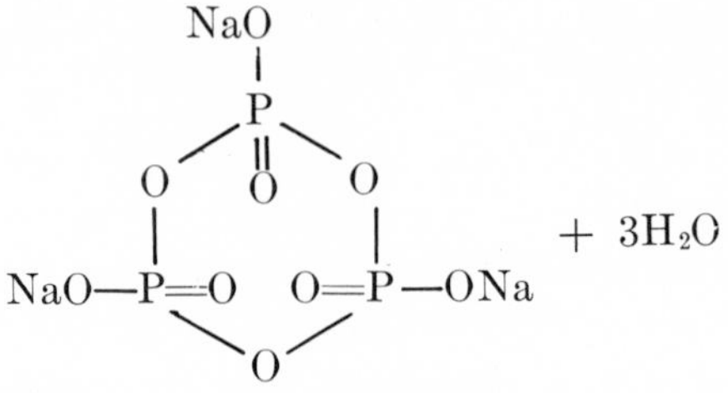

cyclic trimetaphosphate

The product is a white crystalline compound called cyclic trimetaphosphate, meaning that there are three metaphosphate ($NaOPO_2$) units joined in a cyclic ring. Sodium trimetaphosphate can also be prepared from glassy sodium polymetaphosphate (Chapter 11); the glassy material is heated or tempered at 520°C. It devitrifies, and the long chains break into three metaphosphate unit chains and join into rings.

Under the conditions described, cyclic sodium trimetaphosphate is thermodynamically the more stable form. This means that it requires less effort for the metaphosphate units to arrange themselves into the cyclic forms under these conditions rather than stay in a chain.

Cyclic sodium trimetaphosphate reacts with an alkali such as sodium hydroxide in a water solution to form sodium tripolyphosphate:

$$(NaOPO_2)_3 + 2\,NaOH \longrightarrow$$

cyclic sodium
trimetaphosphate

sodium
tripolyphosphate

Attempts have been made to produce sodium tripolyphosphate *via* this route for detergent compositions, but this process has yet to become commercially important.

Instant Cooking Cereal. DISODIUM PHOSPHATE. Instant hot cereals also contain disodium phosphate. Years ago, it took many hours to cook breakfast cereal, especially products such as farina, to a soft, smooth-tasting mush. Addition of about 1% disodium phosphate to the farina lowers the cooking time to a few minutes. Cereals require less cooking when the pH is brought slightly above 7, and disodium phosphate does that. (If too much is added, resulting in too high a pH, the cereal disintegrates. A strong alkaline solution can actually dissolve starch in the cereal.) Of course, the pH can be adjusted with such bases as sodium carbonate, but the latter would impart a soapy taste to the cereal. Disodium phosphate is, of course, added in advance to instant cereal. One patented composition consists of a mixture of disodium phosphate, along with calcium and iron phosphates as mineral supplements. These ingredients are agglomerated into small particles similar to farina. Young homemakers who have never had to wait for the old-fashioned farina to cook probably think that instant cooking is something that we have always had.

Potato Processing. SODIUM ACID PYROPHOSPHATE. Among its other uses, sodium acid pyrophosphate is also used to treat potatoes by taking advantage of the sequestering property of its pyrophosphate portion. When a white potato is boiled at home, its end often turns grayish-black. The same thing can happen with home-made French fried potatoes. Yet pre-cooked (or frozen) French fries, diced potatoes, canned boiled potatoes, are all bright and white. Commercial potato products have either been dipped or blanched with a 1–2% solution of sodium acid pyrophosphate to keep their pearly color.

Food chemists believe that after-cooking darkening of potatoes is caused by iron compounds. The potato plant absorbs iron from the soil and deposits most of it as a colorless, ferrous (Fe^{2+}) organic complex compound in the stem end of the tuber. During cooking the iron is freed from the organic complex, and it can combine with tannin-like compounds in the potato. On exposure to air this iron–tannin complex oxidizes to deeply colored ferric (Fe^{3+}) compounds. The pyrophosphate ion sequesters the iron as a colorless iron pyrophosphate complex—*i.e.*, a compound held together by both strong ionic bonds and electrostatic forces; it thus prevents the iron from reacting with the tannin-like material later on to form dark compounds.

Untreated potatoes show extreme after-cooking darkening caused by the release of iron compounds.

These potatoes were boiled in water containing 2% SAPP (sodium acid pyrophosphate). Sapp chelates the iron.

SAPP prevents cooked sweet potatoes from darkening by the same complexing reaction. For processed sweet potato products (frozen, dehydrated, and pureed) the natural color can be maintained without darkening by adding 0.3–0.4% sodium acid pyrophosphate, alone or in combination with tetrasodium pyrophosphate, to the potato during processing. For obvious reasons one of the tradenames for the sodium acid pyrophosphate used for potato treatment is Taterfos.

Tricalcium Phosphate as Anti-Caking Agent

When we shake salt onto our food it usually flows freely, even in humid weather. Sodium chloride, the common table salt obtained from mines or from the sea, even after many purification steps still contains a small amount of magnesium chloride as an impurity. Magnesium chloride which is hygroscopic, absorbs moisture and causes the salt to agglomerate. To prevent this, tricalcium phosphate, $Ca_5(PO_4)_3(OH)$, is used. It consists of fine particles, about 1 to 2 microns in size (one micron is 1×10^{-6} meter), which coat the larger salt crystals and prevent sticking.

Tricalcium phosphate is used also to impart free-flowing properties to other powders—*e.g.*, granulated sugar, baking powders, and even fertilizers. Usually the addition of 1% is sufficient, and seldom more than 2–3% is needed.

Most people use the formula $Ca_3(PO_4)_2$ for tricalcium phosphate. Actually, the industrial tricalcium phosphate which is used for conditioning and other applications has a composition which corresponds to hydroxyapatite: $Ca_5(PO_4)_3(OH)$. Hydroxyapatite is prepared commercially by adding phosphoric acid to a slurry of hydrated lime. This order of addition of reactants is just the reverse of that for preparing the monocalcium and dicalcium phosphates. In this case the aim is to neutralize all the hydrogen ions in the phosphoric acid, and with this order of addition, excess lime is always present. The hydroxyapatite formed is extremely insoluble and precipitates immediately as fine particles, from less than a micron to just a few microns in diameter. The precipitate is centrifuged from the mother liquor, dried, and milled to go through a 325-mesh (less than 0.0018 inch diameter) screen.

5

Dentifrices

Dicalcium Phosphates

Dicalcium Phosphate Dihydrate (CaHPO₄·2H₂O). Most toothpastes contain a phosphate as a polishing agent, usually dicalcium phosphate dihydrate (if the toothpaste does not contain fluoride). A minor percentage of pastes uses silica. Of the 60–70 million pounds of crystalline dicalcium phosphate dihydrate produced annually in the United States, about two-thirds winds up in toothpaste.

Dicalcium phosphate dihydrate is prepared industrially by adding a dilute slurry of hydrated lime [$Ca(OH)_2$], a base, to a 30–40% water solution of phosphoric acid. In this reaction two hydrogen atoms in phosphoric acid are replaced by one calcium atom. The reaction is controlled by addition of the correct ratio of lime to phosphoric acid; to ensure that only the dihydrate (and no anhydrous material) is formed, the temperature is kept at 40°C or lower. The reaction is:

$$H_3PO_4 + Ca(OH)_2 \longrightarrow CaHPO_4 \cdot 2H_2O$$

Dicalcium phosphate dihydrate precipitates out and is separated from the mother liquor by centrifuging. It is dried and milled to such fine particles that 99.5% of it will pass through a 325-mesh screen (325-mesh particles have a diameter less than 0.0018 inch).

PREVENTION OF DEHYDRATION. Before dicalcium phosphate dihydrate can be used in a dentifrice, it must first be kept from dehydrating to the anhydrous form after it is formulated into toothpaste. Dehydration results in relatively large crystals of anhydrous dicalcium phosphate which makes the paste gritty, or worse, may cause it to harden like cement in the tube. To stabilize the formulation, a pyrophosphate ion is added until its concentration in the finished product is about 1%. At present we have no good explanation as to how this process works.

The oldest method of stabilization is to add 2–3% trimagnesium phosphate to the finished product. Again, we have no clear-cut explanation as to why trimagnesium phosphate acts as a stabilizer. Minor modifications in additives and stabilizers are necessary to fulfill special requirements desired by particular toothpaste manufacturers. For example, dicalcium phosphate dihydrate stabilized with trimagnesium phosphate is not generally used in toothpastes containing the surfactant (*i.e.,* a cleaning agent like soap) and cavity preventative, sodium lauryl sarcosinate. The magnesium ion in trimagnesium phosphate would react with sodium lauryl sarcosinate to precipitate a crystalline magnesium salt and give the toothpaste a gritty consistency.

DEHYDRATION AND DECOMPOSITION. Dehydration in the case of dicalcium phosphate dihydrate (DPD) would mean the loss of two moles of water to form the anhydrous product shown below:

$$CaHPO_4 \cdot 2H_2O \longrightarrow CaHPO_4 + 2H_2O$$

Actually the chemistry of this particular dehydration reaction is very complex and not well understood. We do know that the reaction is catalyzed by moisture, acid, and unknown impurities; heat also helps. Thus chemists find it surprising that DPD dehydrates faster in a moist atmosphere than when it is kept dry. It sounds paradoxical to say that to keep it wet (hydrated), you must keep it dry.

At lower temperatures, the dehydration reaction yields anhydrous dicalcium phosphate. At higher temperatures, and especially in the presence of hot or boiling water, the reaction goes as follows. First dicalcium phosphate dihydrate hydrolyzes to the more basic octacalcium phosphate $Ca_8H_2(PO_4)_6 \cdot 5H_2O$ and free phosphoric acid:

$$8CaHPO_4 \cdot 2H_2O \longrightarrow Ca_8H_2(PO_4)_6 \cdot 5H_2O + 2H_3PO_4 + 11H_2O$$

As more heat is added, octacalcium phosphate hydrolyzes to the still more basic hydroxyapatite [tricalcium phosphate, $Ca_5(PO_4)_3OH$] and more free phosphoric acid:

$$5Ca_8H_2(PO_4)_6 \cdot 5H_2O \longrightarrow 8Ca_5(PO_4)_3OH + 6H_3PO_4 + 17H_2O$$

This last hydrolysis step has been found empirically to be inhibited by magnesium ions. The hydroxyapatite and phosphoric acid formed from the last step of hydrolysis could combine to produce anhydrous dicalcium phosphate. Since magnesium ions inhibit the hydrolysis, it is assumed that this indirectly prevents the formation of anhydrous dicalcium phosphate. Thus, even the apparently simple removal of water molecules can be quite complex. For toothpaste production, such dehydration must be prevented.

Since we still don't know how the stabilizers work, you may wonder how they were discovered. Guy MacDonald discovered the trimagnesium phosphate stabilizer by trying many and various compounds as additives and found that trimagnesium phosphate worked. Theoretically, this chemistry is still a great challenge. Considerable research is probably furrowing the brows of chemists in various laboratories as they strive to unravel the riddle of missing water.

USE AS DENTIFRICE. Dicalcium phosphate dihydrate made its debut on toothbrushes in the early 1930's, replacing the then commonly used precipitated chalk in toothpastes. A dicalcium phosphate dihydrate dentifrice not only cleans teeth, but it is less abrasive than chalk and puts a glossier shine on tooth enamel. This advantage is endlessly touted on television and in magazines and newspapers, usually with a picture of a beautiful smiling girl with pretty teeth. Despite its virtues as a dentifrice, however, it cannot be used in toothpastes containing fluoride ions; although it is relatively insoluble in water, sufficient solubility remains to produce enough calcium ions to remove all of the fluoride ions as a precipitate of the extremely insoluble calcium fluoride (CaF_2). Unfortunately, insoluble fluoride is unavailable to the teeth.

The formulation of a toothpaste—more elegantly known as dental cream—requires a polishing agent, a humectant (to keep the paste from drying out), an organic surfactant (for cleansing and foaming), a binder (to keep ingredients from separating and to add body to the paste), a preservative (to prevent bacteria and mold growth), a sweetener, flavoring, and water. A typical formulation is shown below.

		% by weight
Polishing agent:	dicalcium phosphate dihydrate	50
Humectant:	glycerine (95%)	30
Organic surfactant:	sodium lauryl sulfate	1
Binder:	carrageenan gum	1
Sweetener:	saccharin, soluble	0.1
Preservative:	ethyl parasept	0.1
Flavoring:	peppermint	1
Distilled water:		16.8

Note the major role of dicalcium phosphate dihydrate. Considering how much toothpaste we use every year it is little wonder that so many millions of pounds of dicalcium phosphate dihydrate are consumed every year.

Anhydrous Dicalcium Phosphate ($CaHPO_4$). Anhydrous dicalcium phosphate can be manufactured in the same equipment used to make the dihydrate; the only difference is that the reaction is carried out above 70°C to eliminate formation of the dihydrate.

Anhydrous dicalcium phosphate competes with the hydrated version as a polishing agent in toothpaste. However, it is so much more abrasive than dicalcium phosphate dihydrate that if used alone as the polishing agent, it would wear the teeth away to nothing. Thus, small amounts of it are used in combination with the dihydrate to remove stains, for example, from the teeth in specialty items such as "smoker's toothpaste." It is also combined with less abrasive polishing agents such as tricalcium phosphate. Although the market for anhydrous dicalcium phosphate is still relatively small, it could be larger if someone were to discover a process for controlling its abrasiveness. Research in this area is receiving special attention in many dentifrice laboratories.

Calcium Pyrophosphate. PREPARATION. The first important fluoride toothpaste was made with calcium pyrophosphate, the most insoluble and inert of all calcium phosphates. This compound is prepared by the thermal dehydration of dicalcium phosphate dihydrate:

$$CaHPO_4 \cdot 2H_2O \longrightarrow Ca_2P_2O_7 + 3H_2O$$

When this dehydration reaction is carried out under carefully controlled conditions in a humid atmosphere, water is lost rapidly at $120°-160°C$, corresponding almost exactly to the theoretical value for complete removal of the two molecules of water of hydration. Upon further heating, at above $400°C$, anhydrous dicalcium phosphate begins to lose water and condenses into calcium pyrophosphate. (There are several crystalline forms of calcium pyrophosphate. The gamma form is made at around $530°C$, the beta and alpha forms are made at successively higher temperatures.)

Because of its chemical inertness to fluoride ions, extremely low solubility in water, and proper abrasiveness, calcium pyrophosphate was the first polishing agent found compatible for use in fluoride toothpastes. Dental researchers have known for years that a minute amount of fluoride ion strengthens teeth and helps to prevent cavities, especially when teeth are being formed in children. Accordingly, addition of fluoride to toothpaste is an effective means of cavity prevention.

However, a soluble fluoride cannot be used with toothpaste formulations containing dicalcium phosphate dihydrate as the polishing agent. As discussed earlier, even though dicalcium phosphate dihydrate is quite inert and insoluble, it still is sufficiently soluble in the toothpaste mix so that the calcium ion (Ca^{2+}) will precipitate the fluoride ion (F^-) as the extremely insoluble calcium fluoride (CaF_2); thus, it can no longer react with tooth enamel to prevent cavities.

The first fluoride-containing toothpaste with calcium pyrophosphate as polishing agent was Procter & Gamble's Crest. Research leading to the formulation of Crest indicated that stannous fluoride, SnF_2, in combination with stannous pyrophosphate, $Sn_2P_2O_7$, is effective in preventing cavities. Calcium pyrophosphate was chosen as the polishing agent. In the book, "Accepted Dental Remedies," issued by the American Dental Association in 1960, Crest toothpaste was listed as a dentifrice containing 0.4% stannous fluoride, 39% calcium pyrophosphate, 10% glycerin, 20% sorbitol (70% solution), 1% stannous pyrophosphate, 24.9% water, and 4.63% miscellaneous formulating agents.

The calcium pyrophosphate used in toothpaste is actually a mixture of its gamma and beta forms. From the standpoint of chemical inertness, the pure beta and alpha forms prepared at the higher temperatures are ideal, but they are too abrasive for teeth; the selected mixture is a good compromise which is sufficiently inert and is abrasive enough but not too abrasive.

METHODS FOR ABRASIVENESS DETERMINATION. One of the most common methods used to measure the abrasiveness of dental polishing agents was developed by the American Dental Association (ADA). A specially constructed machine holds standardized toothbrushes against antimony metal strips. The polishing agent to be tested is formulated into a standard toothpaste. After the antimony strip is brushed mechanically with this paste for 10,000 back-and-forth strokes, the antimony strip is washed and weighed. The abrasiveness of the polishing agent is calculated by measuring how much antimony disappears from the strip—in other words, its weight loss. This test assumes that what is too abrasive for antimony would also be too abrasive for teeth. Typical abrasive values for various polishing agents obtained by this method are listed below:

Polishing Agent	*ADA Method*
Dicalcium phosphate dihydrate	2–3
Dicalcium phosphate anhydrous	30–50
Hydroxyapatite	2
Calcium pyrophosphate	10–12
Insoluble sodium metaphosphate	2–3

Another commonly accepted method for determining abrasiveness is the RDA procedure (radioactive dentin abrasion). Extracted human teeth are cut to remove the enamel-covered portions. The remaining root portions, known as dentin, are irradiated in a nuclear reactor. These irradiated dentin samples, after aging for 7 to 10 days, are polished with a slurry of the polishing agent under standardized conditions. In this process, some portion of the dentin is removed, and the amount removed is measured by the radioactivity of the slurry. The more abrasive the polishing agent, the more dentin is removed. All tests are run against a control which is a standard sample of calcium pyrophosphate having a RDA value of 500. Less abrasive polishes have RDA values less than 500; more abrasive ones are greater than 500. The desired value is around 500.

Machine for testing the abrasiveness of dentifrices using the antimony metal strip method

Machine for testing abrasiveness of dentifrices using the radioactive dentin (RDA) method

Insoluble Sodium Metaphosphate

Insoluble sodium metaphosphate is also used as a polishing agent in toothpaste. It is especially suited to fluoride-containing anti-cavity toothpastes. Even though the IMP is slightly soluble in water, the positive cation liberated is the sodium ion. Thus, when it reacts with the fluoride, it only forms sodium fluoride, which is still soluble and reactive with the teeth. IMP-containing fluoride toothpaste has gained acceptance. It offers, in addition to the fluoride ions, good polishing and cleansing action.

6

Fertilizers

Phosphoric acid is usually sold and shipped as a clear, easy-to-handle liquid in 75%, 80% or 85% concentrations. The remaining 25%, 20% and 15% is water. A 100% phosphoric acid concentration tends to crystallize to a solid which melts at 42.35°C. This solid is hygroscopic; that is, it grabs water from the air and becomes mushy. In shipping dilute phosphoric acid, especially for large scale operations, the 15% to 25% water represents a major expense. Obviously a better way of handling higher concentrations of this is needed. This need has resulted in the development of superphosphoric acid.

Superphosphoric Acid

Various investigators, especially at the Tennessee Valley Authority where fertilizer research is an important project, studied the liquid–solid phase equilibrium for concentrated phosphoric acid to determine the solidification temperature of liquid phosphoric acids as the phosphorus pentoxide concentration increases. Figure 1 shows what happens. The solidification temperature, or eutectic point, for a 75–76% P_2O_5 acid is lower than that for a 72.6% P_2O_5 concentration (equivalent to 100% H_3PO_4), and about the same as that for dilute 60% P_2O_5 (equivalent to 82% H_3PO_4). Since 75–76% phosphoric acid is the eutectic, it can be shipped and handled as a liquid. Phosphoric acid at this concentration is called superphosphoric acid. It contains orthophosphoric acid, about 42% pyrophosphoric acid, and some tripolyphosphoric acid.

When the pyrophosphoric acid and tripolyphosphoric acid in superphosphoric acid are diluted with water and heated, they revert to orthophosphoric acid. One hundred pounds of superphosphoric acid diluted with water gives about 105 lbs of 72.6% P_2O_5 phosphoric acid. Since the 72.6% P_2O_5 phosphoric acid is the 100% H_3PO_4, it is general practice to say that superphosphoric acid is equivalent to 105% H_3PO_4. Superphosphoric is used as a source of orthophosphoric acid for many applications —e.g., for augmenting the phosphoric acid content of a bright dip bath for chemically polishing aluminum without adding water. Superphosphoric acid prepared from thermal phosphoric acid is also used to prepare liquid fertilizers, although not as extensively as superphosphoric acid prepared from the cheaper wet-process phosphoric acid.

Wet process superphosphoric acid contains considerable metallic impurities such as iron and aluminum, and its P_2O_5 content is around 70%. For liquid fertilizer, the pyrophosphoric acid content is most important. This compound holds the metal impurities in solution through a complexing action known as sequestering. If pyrophosphate were absent, the metallic impurities would precipitate out and foul the equipment. The high concentration of P_2O_5 in superphosphoric acid also permits the manufacture of fertilizers with very high phosphorus content.

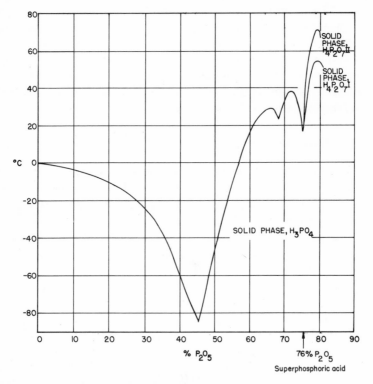

Figure 1.   Liquid–solid phase diagram for phosphoric acid systems

Wet process superphosphoric acid can be used directly as fertilizer. In one method it is added directly to irrigation water.

Calcium Phosphates

Phosphorus is essential for plant growth. It is available from phosphorus-containing plant foods—*i.e.*, from fertilizers, or from finely ground phosphate rock. However, the low solubility of phosphate rock (especially of the fluorapatite) restricts its availability to plants.

Superphosphate. The idea of converting insoluble tricalcium phosphate or phosphate rocks into soluble monocalcium phosphate dates back to 1830. At that time the German chemist, Justus Liebig reported that acidulated bones (monocalcium phosphate) made a good fertilizer. By the 1870's, acidulated phosphate rocks were produced industrially to be used as fertilizer. Today, superphosphate is made by the reaction of accurately proportioned, finely ground phosphate rock with sulfuric acid. Tricalcium phosphate in phosphate rock is converted to hydrated monocalcium phosphate. Calcium sulfate, the other product, is mixed in, and the hydrogen fluoride leaves. The simplified reaction is:

$$2\ Ca_5(PO_4)_3F + 7\ H_2SO_4 + H_2O \rightarrow 7\ CaSO_4 + 3\ Ca(H_2PO_4)_2 \cdot H_2O + 2\ HF$$

This reaction is exothermic, and the liberated heat drives off a large amount of water as steam. To ensure that the product is free of uncombined phosphoric acid or sulfuric acid (either of which would make the product hygroscopic and cause caking), the amount of sulfuric acid added is less than that theoretically called for. As a result, a small quantity of dicalcium phosphate dihydrate appears in the product along with some undecomposed phosphate rock.

The reaction mixture is cured by storage, sometimes for up to three weeks, to complete the reaction. After curing and drying, the fertilizer is ground and bagged.

In the original reaction and during the curing some fluoride in phosphate rock evolves as hydrogen fluoride gas. Some of it comes off as silicon tetrafluoride (SiF_4) from the reaction of hydrogen fluoride with silica. However, about 50–75% of the fluoride remains in the fertilizer product, which also contains 18–20% available P_2O_5.

Triple Superphosphate. In superphosphate fertilizer, the presence of calcium sulfate by-product acts only as an inert diluent; it has almost no nutrient value for plants. A more desirable product is monocalcium phosphate monohydrate which contains very little useless calcium sulfate. The desire for this product led to the development of "triple superphosphate" fertilizer. Triple superphosphate is prepared by the decomposition of phosphate rock with phosphoric rather than sulfuric acid:

$$Ca_5(PO_4)_3F + 7\ H_3PO_4 + 5\ H_2O \rightarrow 5\ Ca(H_2PO_4)_2 \cdot H_2O + HF$$

Thermal phosphoric acid (74–78% concentration) can be used, but it is generally too expensive for this purpose. Instead, wet process acid is mixed with finely ground phosphate rock in correct proportions to produce monocalcium phosphate. The thin slurry discharges into the upper end of a kiln that is lined with firebrick and heated with gas or oil. The

Triple superphosphate granules of different analyses and sizes are kept in separate piles inside this bulk storage building. Shuttle belt conveyors overhead distribute product to the proper pile. A front-end loader reclaims for bagging or bulk shipment.

hot mixture, still in the form of a slurry, is then dumped into a concrete bin (called a den) where it hardens into a porous mass. After aging to complete the reaction, the product is dried of residual moisture and is then ready for packaging and shipping. Triple superphosphate contains about 45–50% available (water-soluble) P_2O_5.

Ammonium Phosphates

Ammonium phosphates which contain both nitrogen and phosphorus needed by plants are often used in combination with potassium phosphates. These combination fertilizers which promote root and leaf development, lead to better flower development and better fruit production.

The ammonium salts of superphosphoric acid have a higher concentration of ammonia and phosphorus than diammonium phosphate and are favorites as fertilizers. They are used either in solid form or in solutions.

Potassium Phosphates

Potassium phosphates are valuable compounds for special applications—*e.g.*, in the catalyst system for synthetic rubber manufacture; for radiator coolants; as a soluble fertilizer; and for liquid detergents. Like their sodium counterparts, potassium orthophosphates are prepared by replacing the hydrogen atoms in phosphoric acid with potassium atoms. Typical reactions are:

$$H_3PO_4 + KOH \longrightarrow KH_2PO_4 + H_2O$$

phosphoric   potassium   mono-
     acid   hydroxide   potassium
                               phosphate

$$H_3PO_4 + 2\ KOH \longrightarrow K_2HPO_4 + 2\ H_2O$$

                               dipotassium
                               phosphate

$$H_3PO_4 + 3\ KOH \longrightarrow K_3PO_4 + 3\ H_2O$$

                               tripotassium
                               phosphate

The preferred source for sodium in the preparation of sodium phosphates is sodium carbonate (soda ash). The more expensive sodium hydroxide is used only when sodium carbonate cannot be used for chemical reasons. The above equations show that the opposite is true for potassium. Potassium hydroxide is preferred over potassium carbonate as the source of potassium because the former happens to be cheaper than the latter.

Some of the industrially available potassium phosphates are:

KH_2PO_4	Monopotassium phosphate
K_2HPO_4	Dipotassium phosphate
K_3PO_4	Tripotassium phosphate
$K_4P_2O_7$	Tetrapotassium pyrophosphate
$K_5P_3O_{10}$	Potassium tripolyphosphate
$(KPO_3)_n$	Potassium polymetaphosphate

Because of their higher cost, the production of potassium phosphates is limited. For example, annual production of sodium tripolyphosphate is over 2.2 billion lbs, but yearly production of potassium tripolyphosphate is practically nil. Nevertheless, potassium phosphates have unique properties which are needed in special applications.

Potassium Orthophosphates. MONOPOTASSIUM PHOSPHATE. Monopotassium phosphate is prepared industrially by adding one mole of a water solution of potassium hydroxide to one mole of phosphoric acid. The solution is concentrated by boiling off water. Upon cooling, crystals of monopotassium phosphate separate and are removed by centrifuging. The mother liquor is recycled for the next batch. At 30°C, the solubility of monopotassium phosphate is a little over 20 grams per 100 grams of solution.

Manufacturers of everything from gin and vodka to antibiotic wonder drugs use monopotassium phosphate as a mineral nutrient for the microbes in their fermentation tanks. For the rapid growth of microbes (molds, yeasts and bacteria), both potassium and phosphorus are usually essential. The organisms use the phosphate in energy transfer reactions. For example, in the fermentation of glucose (a simple sugar), glucose 1-phosphate is the first compound formed. Depending on the nature of the yeast used, this can lead finally to the formation of alcohol or other fermentation products. Experiments have shown that the rate of the fermentation can be gauged by the rate of orthophosphate disappearance. Phosphate plays such an important role in all biological processes that if it is absent from the culture medium, the microbes simply won't grow.

Phosphorus is essential to plant growth as shown by the fertilized plant on the left vs. the untreated plant on the right

Since monopotassium phosphate goes readily into solution, it is ideal for liquid fertilizer. A special commercial fertilizer formulation ($KH_2PO_4 \cdot (NH_4)_2HPO_4$) consists of equal parts of monopotassium phosphate and diammonium phosphate. It contains about 10.5% nitrogen, 52% P_2O_5, and 17.5% K_2O (potassium oxide) and thus is rated as 10-52-17. (On commercial fertilizer packages the three consecutive numbers such as 10-10-10 or 15-6-8 represent the $N–P_2O_5–K_2O$ content.) Dried cow manure is 2-1-2. This potassium-ammonium phosphate mixture is very soluble in water and extremely effective in minimizing shock when seedlings such as tomatoes, tobacco, peppers, and cabbages are transplanted. The shock from transplanting usually causes the plants to wilt and sometimes die. It takes several days to resume their normal growth.

7

Phosphates for Cleaners, Detergents, and Dispersants

The most widely used phosphates for cleaning and detergent compositions are the sodium phosphates. They are also very important in
metal cleaning, water softening, various food processing, and toothpastes.
As a class, however, they represent many compounds. Sodium orthophosphates are sodium salts of phosphoric acids. Hydrogen atoms in
phosphoric acids are replaced by sodium atoms. Sodium orthophosphates
are prepared by successively replacing the hydrogen atoms in orthophosphoric acid by one, two, or three sodium atoms. For example, the reaction
below shows the formation of monosodium phosphate:

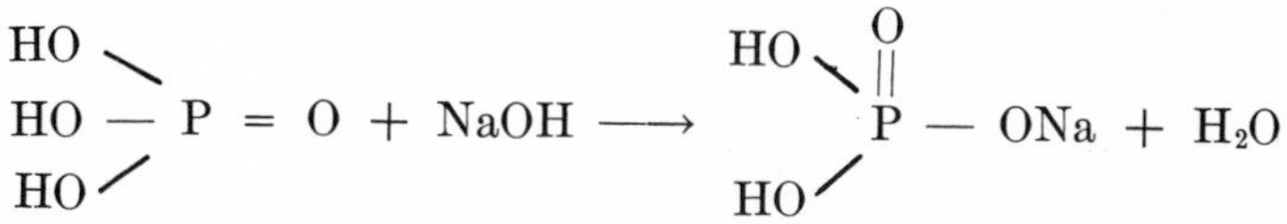

$$\begin{array}{c} \text{HO} \\ \text{HO} - \text{P} = \text{O} + \text{NaOH} \longrightarrow \\ \text{HO} \end{array} \quad \begin{array}{c} \text{HO} \quad \overset{\text{O}}{\underset{\|}{}} \\ \text{P} - \text{ONa} + \text{H}_2\text{O} \\ \text{HO} \end{array}$$

Sodium phosphates containing more than one phosphorus atom in
a single molecule—as in sodium pyrophosphate (or sodium diphosphate)
and sodium tripolyphosphate—can also be made by replacing the hydrogen atoms in the corresponding pyrophosphoric acid or tripolyphosphoric acid with sodium atoms. Commercially, this class of sodium polyphosphates is not manufactured by this route. Their preparative pro-
cedures are described under their respective sections.

Many sodium phosphates form hydrates. Some sodium phosphates
are available in both anhydrous and various hydrated forms. Since there
are so many commercially important sodium phosphates, the list on
page 70 will help differentiate them.

Twelve different sodium phosphates are listed. Even though they
are all from the same family, each is unique, and sometimes they behave
as though they are not at all related. Chemists take advantage of these
differences and use each compound for quite different duties depending
on the individual talents of each.

Trisodium Phosphate for Cleaning Painted Surfaces

Many forms of trisodium phosphate are commercially available. One
is the anhydrous form, Na_3PO_4, in which all three of hydrogens in phos-

List of Sodium Phosphates

NaH_2PO_4	Monosodium phosphate
Na_2HPO_4	Disodium phosphate
Na_3PO_4	Trisodium phosphate
$(Na_3PO_4 \cdot 12\,H_2O)_5 \cdot NaOH$	Trisodium phosphate dodecahydrate–sodium hydroxide complex
$(Na_3PO_4 \cdot 11\,H_2O)_4 \cdot NaOCl$	Trisodium phosphate–sodium hypochlorite complex
$Na_2H_2P_2O_7$	Sodium acid pyrophosphate
$Na_4P_2O_7$	Tetrasodium pyrophosphate
$Na_5P_3O_{10}$	Sodium tripolyphosphate
$-(NaPO_3)_{15\text{-}20}-$	Glassy sodium polymetaphosphate
$-(NaPO_3)_n-$	Insoluble sodium metaphosphate
$(NaPO_3)_3$	Cyclic trimetaphosphate
$(NaPO_3)_4$	Cyclic tetrametaphosphate

phoric acid have been replaced by sodium atoms. Another is trisodium phosphate dodecahydrate. It also contains about one mole of sodium hydroxide for every five moles of trisodium phosphate $[(Na_3PO_4 \cdot 12\,H_2O)_5NaOH]$. Trisodium phosphate, or TSP, is the compound purchased for home use.

In preparing trisodium phosphate, sodium atoms from sodium carbonate replace hydrogen atoms from phosphoric acid. However, only about two of the hydrogen atoms in phosphoric acid can be replaced this way. Sodium carbonate is not basic enough to replace the third. In commercial manufacturing practice, the third hydrogen atom in phosphoric acid is replaced by sodium from sodium hydroxide—a very strong base; thus, about two-thirds of the sodium comes from sodium carbonate and one-third from the more expensive sodium hydroxide.

Trisodium compounds are strongly alkaline, especially the dodecahydrate which contains the extra sodium hydroxide. Many trisodium phosphate applications depend on this high alkalinity. For example, in cleaning painted walls with trisodium phosphate, the alkalinity of the solution causes a thin layer of paint which has been oxidized by air to be removed from the wall surface. The clean underlayer is thus exposed. If a hot solution containing about 10% trisodium phosphate is used, all of the painted coating is removed from wooden or metal surfaces. The high alkalinity of trisodium phosphate saponifies (splits apart) the compounds (fatty acid esters) which hold the oil-based paint together.

In many ways, paint removal is closely related to soap making since fats are broken up in both processes. Chemically, fat is a glyceride—*i.e.*, a compound of glycerine and high molecular weight organic acids; these acids are often called fatty acids. Their nature varies depending on the type of fat from which they are derived. For example, fatty acids of beef tallow are mostly stearic acid with 18 carbon atoms, palmitic acid with 16 carbon atoms, and oleic acid with 18 carbon atoms and a double bond. The following reaction shows the saponification of a fatty acid ester (glyceride) into glycerine and the sodium salt of palmitic acid (soap) with trisodium phosphate:

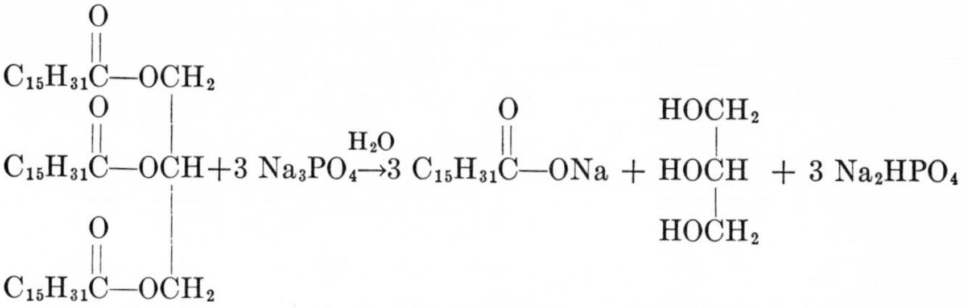

$$
\begin{array}{c}
\overset{\text{O}}{\underset{\|}{}} \\
C_{15}H_{31}\overset{\|}{C}\!-\!OCH_2 \\
\end{array}
$$

$$
\begin{array}{ccccc}
\overset{\text{O}}{\underset{\|}{}} & & \overset{\text{O}}{\underset{\|}{}} & & HOCH_2 \\
C_{15}H_{31}\overset{\|}{C}\!-\!OCH + 3\ Na_3PO_4 \xrightarrow{H_2O} & 3\ C_{15}H_{31}\overset{\|}{C}\!-\!ONa & + & HOCH & +\ 3\ Na_2HPO_4 \\
& & & HOCH_2 & \\
\overset{\text{O}}{\underset{\|}{}} & & & & \\
C_{15}H_{31}\overset{\|}{C}\!-\!OCH_2 & & & &
\end{array}
$$

fat   soap   glycerine
(sodium palmitate)

The ability of trisodium phosphate to break fats and greases into water-soluble glycerine and soap makes it extremely useful in scouring powders. When combined with abrasives and chlorine-generating bleaches, it removes cooking grease and stains.

Trisodium Phosphate–Sodium Hypochlorite Complex for Cleaning, Bleaching, and Sanitizing

In many industries, such as in food processing and dairy plants where cleanliness is important, the high alkalinity of trisodium phosphate makes it the cleansing agent of choice. After cleaning, however, the equipment must still be sterilized. One of the sanitizing agents used is sodium hypochlorite ($NaOCl$); this potent oxidizer and chlorinating compound is prepared by the action of chlorine (Cl_2) on a water solution of sodium hydroxide ($NaOH$):

$$2\ NaOH + Cl_2 \longrightarrow NaOCl + NaCl + H_2O$$
sodium
hypochlorite

Sodium hypochlorite is familiar to us as liquid laundry bleach (Clorox, Purex, etc.). It is stable only in a solution containing excess sodium hydroxide; as a solid, it quickly decomposes. For cleaning and sanitizing food-handling equipment, a one-step process is highly desirable. If trisodium phosphate and sodium hypochlorite could be combined into one compound, cleansing and sanitizing could be done in one operation.

In the early 1920s, L. D. Mathias, a chemist with Victor Chemical Works, working on this problem, theorized that since sodium hypochlorite is prepared by the action of chlorine on sodium hydroxide, and trisodium phosphate dodecahydrate is actually a complex of trisodium phosphate with sodium hydroxide, it might be possible to obtain a reaction between the chlorine and the sodium hydroxide in the trisodium phosphate complex. A new trisodium phosphate–sodium hypochlorite complex would result. The compound he envisioned would have the formula $(Na_3PO_4 \cdot 12\ H_2O)_5NaOCl$, and the reaction would be:

$$(Na_3PO_4 \cdot 12\ H_2O)_5 \cdot NaOH + NaOH + Cl_2 \longrightarrow$$
$$(Na_3PO_4 \cdot 12\ H_2O)_5NaOCl + NaCl + H_2O$$

His hunch was 99% right. After much laboratory work, a new compound with the structure $(Na_3PO_4 \cdot 11\ H_2O)_4NaOCl$ was discovered— not exactly the compound Mathias had planned, but very close. It was given the tradename Cl-TSP for chlorinated trisodium phosphate. One of the processes developed to manufacture it involves the addition of a sodium hydroxide solution of sodium hypochlorite to a concentrated solution of disodium phosphate. Disodium phosphate reacts with sodium hydroxide to form trisodium phosphate in solution. It, in turn, reacts with sodium hypochlorite to form the trisodium phosphate–sodium hypochlorite complex, and then the complex (Cl–TSP) precipitates out as a stable solid.

Subsequent investigators discovered that trisodium phosphate forms a complex not only with sodium hydroxide and sodium hypochlorite but with many other sodium salts. One of the interesting complexes is that formed with sodium permanganate $(Na_3PO_4 \cdot 11\ H_2O)_7NaMnO_4$. Since this complex is deep purple, it was instrumental in eliminating a problem of mistaken identity. Cooks in the kitchens of dining cars on trains had been using Cl–TSP for cleaning and sanitizing equipment. Since Cl–TSP is a white powder like sugar and salt, careless waiters sometimes filled sugar bowls or salt shakers with it by accident. Cl–TSP tastes like soap and when shaken on meat or scrambled eggs produced enraged roars from Pullman passengers.

To prevent such accidents, research came to the rescue. One obvious solution—adding a dye to Cl–TSP—posed problems because sodium hypochlorite in Cl–TSP bleaches most organic dyes. The difficulty was finally solved by adding a small amount of sodium permanganate to the reaction mixture. The entire product is colored by the trisodium phosphate–sodium permanganate complex to an attractive purple-pink, yet the cleaning and sanitizing action of the main component, Cl–TSP, is unaffected.

When mixed with an abrasive, such as a silica, and a detergent such as sodium alkylbenzenesulfonate, Cl–TSP is a very effective cleanser for hard surfaces such as kitchen sinks and bath tubs. The chlorine liberated upon addition of water bleaches the stains almost immediately. The compound is also effective in automatic dishwashing detergents. It is formulated with 40–55% sodium tripolyphosphate and a surfactant, along with sodium metasilicate. Cl–TSP provides high alkalinity which disperses grease. Since it is water soluble, it is easily rinsed from glass surfaces. It provides a sheeting action on glass surfaces which helps remove soil and cleaning formulation components during the washing and rinsing cycles. This action cleans glasses, leaving them streak-free and spotless. The chlorine liberated from the Cl–TSP also bleaches stains from dishes.

Tetrasodium Pyrophosphate (TSPP) in Detergents

Tetrasodium pyrophosphate is made by removing one mole of water from two moles of disodium phosphate:

$$(NaO)_2P{-}OH + HO{-}P(ONa)_2 \xrightarrow{500°C} (NaO)_2P{-}O{-}P(ONa)_2 + H_2O$$

disodium phosphate   tetrasodium
pyrophosphate

This reaction is carried out at about 500°C. The product is a white powder, soluble in water to the extent of about 10%. This relatively low solubility limits its use in some applications, such as in liquid detergents. However, tetrasodium pyrophosphate (TSPP) is still very useful in soaps where the 10% solubility in water is more than enough.

TSPP found its first major use as a builder for soap during the 1930's. Chemically, the term builder means that when TSPP is added to soap it produces a more efficient cleaning agent. The action of each ingredient enhances that of the other. Other builders still used to some extent include soda ash (sodium carbonate, Na_2CO_3), sodium silicates, and trisodium phosphate. As builders, however, they are not quite as effective as tetrasodium pyrophosphate.

From the standpoint of chemistry and physics, the cleaning process is extremely complicated. One reason for the effectiveness of tetrasodium pyrophosphate as a builder is its effect on the "critical micelle" concentration of the detergent. The theory behind the cleaning action of soap (or organic surfactant compounds that do the same job) is that a soap molecule has a long organic chain at one end. The other end has a

—COONa group which in water becomes a $-\overset{\overset{\displaystyle O}{\|}}{C}-O^-$ ion with a negative charge. This end is called the polar group, like the negative pole of an electric battery. At a certain critical concentration in water the soap molecules agglomerate into a micelle—that is, about 60–80 molecules of soap aggregate into a small jelled lump with the organic or non-polar end on the inside and the inorganic, polar end on the outside. The organic groups on the inside dissolve the organic oily and greasy portion of the dirt.

The presence of a builder lowers the concentration of soap or organic surfactant in water necessary for micelle formation; that is, it now takes less than 60–80 soap molecules to aggregate into a micelle. Thus, less soap or surfactant is needed for the same cleaning power because many small micelles clean more efficiently than one large micelle.

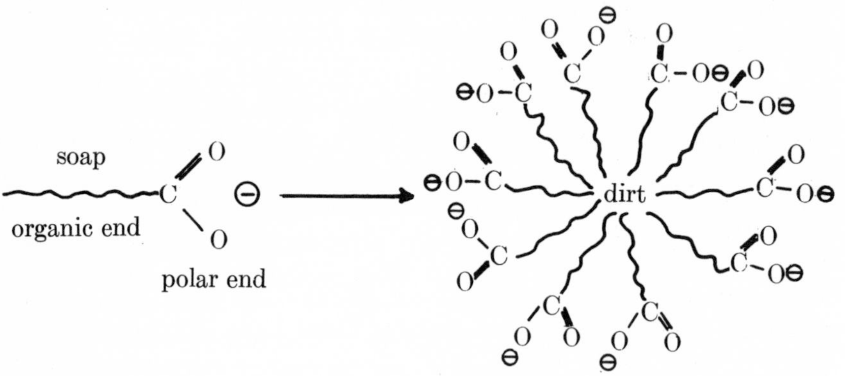

Most dirt particles are attached to fabric by an electrostatic force or by an oily film. In cleaning, the detergent loosens this link so that the particles become suspended in the water. TSPP is a good dispersant, so it helps soap to hold the dirt particles in suspension.

In hard water, as described in Chapter 11 on water softening, calcium and magnesium ions react with the soluble sodium soap to form insoluble calcium and magnesium soaps. These insoluble soaps form a gray scum that can redeposit on the clothing. They also decrease the concentration of the soluble sodium soap in the cleaning solution. TSPP is an effective builder for soap also because at the right concentration it reacts with calcium and magnesium ions to hold them tightly as a complex water-soluble compound; this is called sequestration. The calcium and magnesium ions thus sequestered are no longer available to react with soap to form insoluble calcium and magnesium soaps.

$$2C_{15}H_{31}\overset{\overset{\displaystyle O}{\|}}{C}ONa \ + \ Ca^{++} \ \longrightarrow \ (C_{15}H_{31}\overset{\overset{\displaystyle O}{\|}}{C}O)_2Ca \ + \ 2Na^+$$

Sodium soap   Calcium soap

When TSPP was introduced commercially as a soap builder, sodium tripolyphosphate was not yet available. Since the introduction of synthetic surfactants with sodium tripolyphosphate as the builder, TSPP-built soap has become much less important. One major reason TSPP is no longer used extensively as a commercial sequestering agent is that even though it has a strong complexing action, if present in large excess, it forms extremely insoluble pyrophosphates of Ca^{2+}, Mg^{2+}, and Fe^{2+} if these ions are present in excess. This precipitate formation tends to counteract the complexing ability of the pyrophosphate ion.

Sodium Tripolyphosphate as Detergent Builder

At this writing, sodium tripolyphosphate (STPP) is a much-maligned compound. Its use in detergents has been blamed as the chief cause of pollution which leads to the eutrophication of lakes and streams. As a component in detergents, it has had a profound influence on that industry. On the positive side, it has provided users with a new order of magnitude in the cleanliness of their laundry. "Tattletale gray" shirts and the dull look on colored dresses are now a thing of the past. At present about 2.2 *billion* pounds of sodium tripolyphosphate are produced each year in the United States. Actually this figure is not so staggering when we realize that every U.S. housewife uses sodium tripolyphosphate in her washing machine. Every package of synthetic detergent on the market contains about 25–45% sodium tripolyphosphate, whose primary purpose is as a builder for the synthetic surfactant (detergent), sodium alkylbenzenesulfonate.

Sodium alkylbenzenesulfonate is similar to soap in that it is also a molecule with an organic end and a polar end. The organic end is usually a benzene ring to which is attached an organic alkyl group with an average

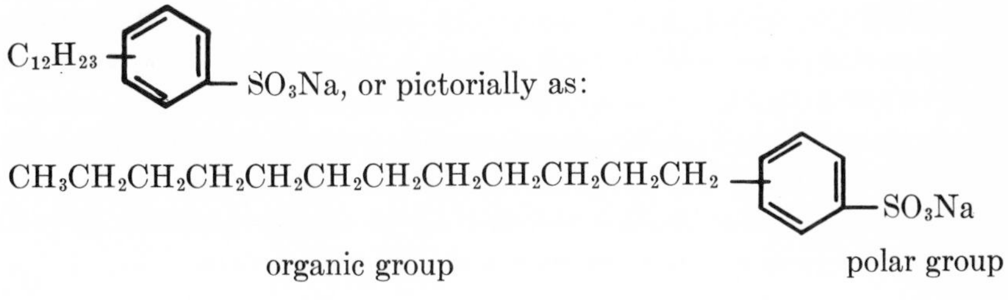

$C_{12}H_{23}$ ─⟨benzene ring⟩─ SO_3Na, or pictorially as:

$CH_3CH_2CH_2CH_2CH_2CH_2CH_2CH_2CH_2CH_2CH_2CH_2$ ─⟨benzene ring⟩─ SO_3Na

organic group   polar group of 12 carbon atoms. The polar end is the sodium salt of a sulfonic acid group ($-SO_3Na$). Sodium alkylbenzene sulfonate can be represented by the formula shown at the bottom of page 75. The alkyl chain attached to the benzene ring can be a straight chain, as illustrated, or a branched chain. The straight chain compounds are now preferred because they can be biologically "chewed up," that is, destroyed in sewage plants by bacterial action. In other words, they are biodegradable, and this property eliminates foaming problems in rivers and streams.

Sodium tripolyphosphate is manufactured by heating a mixture of two moles of disodium phosphate with one mole of monosodium phosphate. Two moles of water are eliminated, as shown below:

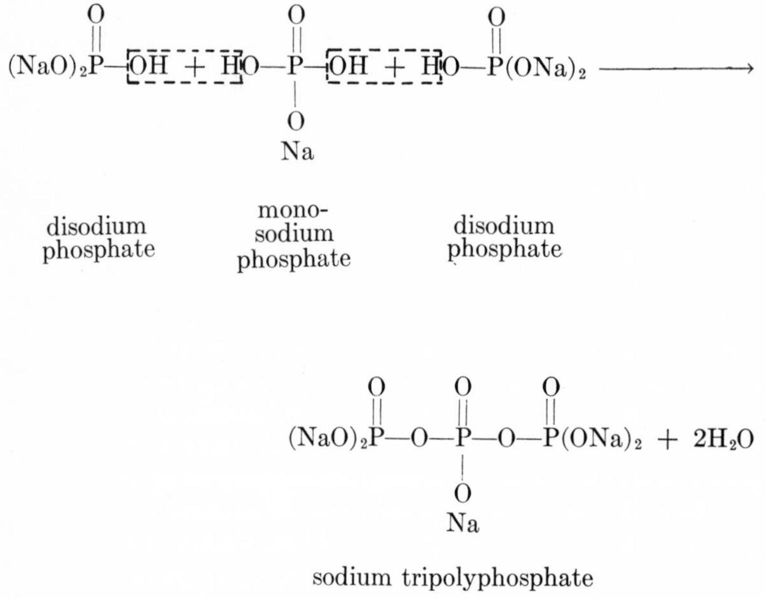

$$(NaO)_2P-OH \; + \; HO-P-OH \; + \; HO-P(ONa)_2 \longrightarrow$$

disodium phosphate	mono- sodium phosphate	disodium phosphate

$$(NaO)_2P-O-P-O-P(ONa)_2 \; + \; 2H_2O \qquad (1)$$

sodium tripolyphosphate

There are two anhydrous forms of sodium tripolyphosphate called Type I and Type II. They are chemically the same, but since their crystal structures are different, they exhibit somewhat different behavior. Type I is formed by carrying out Reaction 1 at 540°–580°C. A pure Type II can be obtained by heating a 1:2 mixture of the mono- and disodium phosphates to a molten state at about 620°C, keeping it at 550°C for some time and then rapidly cooling it to room temperature.

As it cools to 150°–100°C, the glassy product disintegrates into a fine Type II powder. Type I is the stable crystalline phase of sodium tripolyphosphate while Type II is regarded as the metastable phase. Type II can be converted to Type I by heating but not vice-versa.

In industry, a mixture of Types I and II is produced. For those detergent formulations needing a higher percentage of Type I, a temperature range of 415°–490°C is used. When mostly Type II is desired, a temperature range of 350°–400°C is used. At room temperature, commercial grade Type I has a water solubility of about 15–16 grams per 100 grams of solution while Type II has an immediate solubility of about 31 grams per 100 grams of the solution. The higher solubility of Type II is short lived. Within minutes it reacts with water to form the hexahydrate ($Na_5P_3O_{10} \cdot 6\ H_2O$), which has a solubility of only about 14 grams per 100 grams of solution. This conversion to the hexahydrate can also occur with Type I:

$$Na_5P_3O_{10} + 6\ H_2O \longrightarrow Na_5P_3O_{10} \cdot 6\ H_2O \qquad (2)$$
$$\text{sodium tripolyphosphate}$$
$$\text{hexahydrate}$$

Most synthetic detergent producers mix a water slurry of solid sodium tripolyphosphate with sodium alkylbenzenesulfonate and other ingredients. Fortunately, with the mixture of Types I and II, sodium tripolyphosphate does not form the low solubility hexahydrate even after a few hours. One can thus obtain a slurry with higher solids content, which is more economical (less water to remove) when the slurry is subsequently spray-dried to the finished detergent composition.

In the spray-drying process, used by all major detergent manufacturers, sodium tripolyphosphate is eventually converted entirely to the hexahydrate. Some of the hexahydrate is degraded to tetrasodium pyrophosphate and a small amount of by-product sodium orthophosphate. This by-product has no practical value as a detergent builder.

For specialty formulations, large, expensive spray-dry equipment is uneconomical for drying detergent mixtures. Thus, a water solution of sodium alkylbenzenesulfonate is sprayed into a mixer containing a porous sodium tripolyphosphate and other ingredients. This porous compound absorbs aqueous sodium alkylbenzenesulfonate solution, and the absorbed water is used to form dry sodium tripolyphosphate hexahydrate. In this way, a solid mix product containing water of crystallization is produced ready for packaging. Hence, it is necessary to make the compound with the correct balance of Type I and Type II and with the correct porosity.

One of the main functions of STPP as a builder is its chelating action—*i.e.*, its ability to sequester ions such as Ca^{2+} and Mg^{2+} (which are responsible for water hardness) into a stable water-soluble complex. One mole of STPP binds one mole of Ca^{2+}. In this way, hard water is softened. One theory for the stability of the STPP–metal complex is that the metal ions not only bond to the ionic oxygen atoms in STPP by ionic force but

Today's shopper has an amazing array of cleaners and detergents from which to choose. Sodium phosphates are used widely in these compounds.

are also held tightly to the double bond oxygens by electrostatic force.

In its other functions as a builder, STPP (1) increases the efficiency of the surfactant, probably by lowering the critical micelle concentration (discussed in the section on TSPP); (2) furnishes the proper alkalinity for cleaning and yet does not burn eyes and sensitive skin; (3) provides resistance to change in alkalinity during washing; (4) because of its high negative charge and its ability to absorb on dirt particles, it builds up negative charges on dirt particles and the repulsion between like charges keeps the dirt in suspension.

Because of its water-softening action and cleansing power, STPP is also used in automatic dishwashing compounds. A typical product may contain as much as 60% sodium tripolyphosphate. The rest consists of sodium metasilicate (as a corrosion inhibitor) and a trisodium phosphate–sodium hypochlorite complex or other chlorine-generating compounds such as potassium dichlorocyanurate (bleaching agent for stain removal). This general phosphate-containing formulation is so effective that thus far very few substitutes or suggestions have been advanced to replace it.

Tetrapotassium Pyrophosphate in Liquid Detergents

Tetrapotassium pyrophosphate is the most important potassium phosphate. An extremely water-soluble compound (187 grams per 100 grams of water at 25°C) it is also very resistant to hydrolysis. These properties, along with the fact that it is a pyrophosphate, make it ideal as a builder for liquid detergents despite its relatively high cost. In contrast, tetrasodium pyrophosphate is only soluble to the extent of 10%, insufficient for a concentrated liquid detergent.

In the 1950's, manufacturers introduced a liquid detergent for heavy-duty laundering. It was originally designed for automatic washing machines because it is easier to measure and pour than powdered forms. Its convenience and popularity make it an important item in the detergent industry today. A typical liquid detergent is about 11–20% organic surfactants, about 20% builders, and 7–10% of a "coupling agent," along with other minor ingredients. (A coupling agent such as sodium or potassium toluenesulfonate causes the organic surfactants and the builder, which normally don't mix, to form a homogeneous solution.) The rest of the formulation (about 50%) is water.

Tetrapotassium pyrophosphate, the builder, is made by heating a solution of dipotassium phosphate to dryness; heating is continued to

$$
\begin{array}{c}
\text{O} \qquad\qquad \text{O} \qquad\qquad\qquad\qquad \text{O} \qquad\quad \text{O} \\
\parallel \qquad\qquad \parallel \qquad\qquad\qquad\qquad \parallel \qquad\quad \parallel \\
(KO)_2P\overline{OH} + \overline{HO}P(OK)_2 \longrightarrow (KO)_2P—O—P(OK)_2 + H_2O \quad (3)
\end{array}
$$

dipotassium   tetrapotassium

phosphate   pyrophosphate

about 400°C until a white powder is obtained. For use in liquid detergents, it can be sold as a dry powder or as a 60% water solution.

Besides its use as a builder for liquid detergents, tetrapotassium pyrophosphate is also used as a component in catalyst systems to polymerize butadiene and styrene to a superior grade of synthetic rubber, sometimes called "cold rubber" (because the polymerization is generally done at around −10°C). Tetrapotassium pyrophosphate, unlike tripotassium phosphate, depends for its function on the pyrophosphate anion which complexes ferrous iron used in the polymerization. The ferrous–pyrophosphate complex, in its interaction with cumene hydroperoxide, acts as the catalyst that permits the polymerization to be carried out at low temperature. Since the active ion is the pyrophosphate ion, tetrasodium pyrophosphate may also be used. However, many of the recipes call specifically for tetrapotassium pyrophosphate.

8

Phosphates in Eutrophication

Phosphate has become a very popular or unpopular subject, depending on your point of view. Some item on it appears almost every day in the news. Ecologists condemn it, and public officials ban it. Some newspaper editors even regard phosphorus as a deadly element.

The main focus of this attention, of course, is the use of phosphate in detergents. In this role it has been targeted as the pollutant which has caused rapid eutrophication of our lakes, streams, and rivers. All over the United States and in many European countries the call to ban phosphate in detergents is sounded every day. The subject has become so emotional that well-meaning and intelligent people find it difficult to discuss in a rational and scientific manner. This chapter presents some scientific facts about eutrophication and the role phosphorus and other elements play in it.

Many people use the terms pollution and eutrophication interchangeably. Here, pollution will mean the act of introducing a contaminant into a normally clean water system, and eutrophication will mean the process of over-fertilizing such bodies of water as streams, ponds, and lakes with nutrients so as to cause a rapid and excessive growth of aquatic plants and algae.

Eutrophication, as a natural geological process, involves a body of water such as a lake, where organic life develops and multiplies over the years; fish, insects, shellfish, bacteria, algae, and various aquatic plants appear and flourish. With time, the lake bottom accumulates the remnants of organic life and other sediment and builds up. As the lake becomes more shallow, the marine life changes character and so do the aquatic plants. Eventually, the lake becomes so shallow that it is a marshland or swamp; finally, it may become dry land. This eutrophication process normally takes thousands of years for a large body of water. Our present concern is that with the inflow of excess rich nutrients, eutrophication is rapidly accelerated. For example, many people believe that the shallow western basin of Lake Erie has eutrophied an equivalent of 15,000 years in the past 50 years.

One of the best indicators of advanced eutrophication is the presence of a great deal of algae. Since most environmentalists and lawmakers regard the rapid and excessive growth or "blooming" of algae as the indicator of eutrophication, I shall also use algae, and the nutrients which supply their growth, in the same context.

Algae are small plants which live suspended and free-floating in our waters. They are probably the most obnoxious organisms in lakes. The most common species are called blue-green algae; some of them are truly blue-green in color, others are pale yellowish-green, and some are actually red. About 2500 species of blue-green algae are known. Under the right conditions (proper nutrients, right temperature, and correct pH), the rapid growth of these microscopic phytoplankton can easily

Courtesy Minneapolis Star

This ugly scene along a bank of Lake Minnetonka, north of Minneapolis, shows dead and dying algae which prohibit any life in these waters. Algae feed off nutrients poured excessively into lakes from cesspools, sewage treatment plants, and run-off from agricultural lands. Just as the algae receives excessive nutrients, so too does it grow excessively, devouring all oxygen within its reach, eventually causing its own death.

result in water so opaque that an underwater swimmer cannot see more than a few inches in front of him. Depending on the prevailing algal species, a lake may look like green pea soup or be almost blood-red. The Green Bay and the Red Sea get their color primarily from the color of the algae present. During periods of heavy growth, some algal species rise to the surface, where wind blows them onto the shore. Bacterial decomposition of the accumulated dead algae creates an unpleasant stench. Much of the dead algal cells settle slowly, always decaying, to the lake bottom where decay continues. This bacterial decay consumes a large quantity of dissolved oxygen, sometimes to such an extent that normal diffusion of oxygen from air to water is not sufficient to replenish the supply. The resulting oxygen depletion kills fish and promotes the growth of weeds and more algae.

Both the scientific and nonscientific communities agree that if the present rate of pollution continues unabated, eutrophication will indeed become very rapid. They do disagree, however, as to the source of the nutrients which promote the algal growth and the most expeditious methods for eliminating or reducing them.

The enrichment of lakes with nutrients is a very complicated process. Many lawmakers believe that algal bloom is a modern phenomenon, especially since the large-scale use of phosphates in detergents began in the late 1940s; they think that a simple way to stop eutrophication is to remove phosphates from detergents. They tend to forget that when human habitation and land cultivation increase, the inflow of decaying organic matter, animal, and human wastes into our waters also increases. As a result algal bloom is much more frequent. Actually, much of our present dilemma can be traced to the invention of the flush toilet in the 1840s, and this problem has accelerated in time with increasing population growth.

To show that periodic algal blooming is not an exclusively modern phenomenon, Prof. Willy Lange of the University of Cincinnati has cited recorded occurrences in biblical times. One interesting example he cited occurred in 1825 in Lake Murtansee in Switzerland. After a heavy rainfall which washed organic matter and nutrients into the lake, the lake suddenly turned red, like blood. The Swiss couldn't explain what they saw, but they recalled that on that day two centuries earlier, they had won a bloody battle against the Burgundians. They thought that the lake's turning red was nature's reminder of the slaughter and superstitiously called the phenomenon "Burgunder Blut," or Burgundian blood. It has now been determined that the red coloration was the result of excess algae growth. When Prof. Lange obtained a sample of the algal specimen, *Oscillatoria rubescens*, from Switzerland and cultured it in his laboratory, he found that it does color the water red. Today's hysteria by some, giving the impression that the noxious algal bloom in our lakes could be eliminated by removing or reducing phosphates in detergents, is not the result of superstition but an oversimplification of a very complex problem.

First, let's look at what makes algae grow. Phosphorus is an element essential to all human, animal, and plant life. All plant and animal cells contain 3–5% phosphorus in both organic and inorganic form. However, though essential, phosphorus is not the only critical element needed for growth. When Dr. H. Clyde Eyster was at Monsanto Research Corp., he and his co-workers showed that besides hydrogen and oxygen, blue-green algae require about 16-17 other essential elements for growth. The ratios of these elements are shown in Figure 1. The sizes of the circles

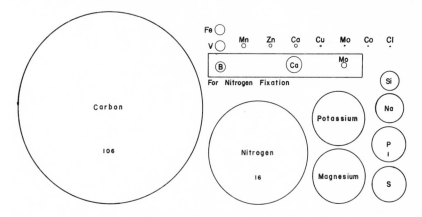

Figure 1.   Ratios of nutrient elements required for algal growth (reproduced by permission of H. Clyde Eyster from his unpublished manuscript: "Algae: Their Total Elemental Nutrient Requirements")

representing the elements show the relative amounts of the various elements needed. Note that the ratio of the number of atoms of carbon to nitrogen to phosphorus needed is approximately 106:16:1. Most of our waters contain all of these elements to a greater or lesser extent. If a lake contains a minimal amount of phosphorus as phosphate, when all of the phosphate is consumed, algae will no longer grow, regardless of the amount of other nutrients present. Thus, phosphorus is the limiting element for algal growth in that lake. The same is true for the other essential elements. In other words, the element which is present in an amount that is depleted first by the growing algae is the limiting element in that particular lake. Some lakes could be nitrogen limiting, and some could be iron or carbon limiting. If carbon is the limiting element, when it is all consumed, no more algae will grow, no matter how much more phosphate is added.

Where do the elements which are required for algal growth come from, and how can we control them? Most of the metallic elements, such as potassium, magnesium, sodium, calcium, and the trace elements (those elements needed for growth in very small amounts) are normally present in the soil and are washed into the lakes. They are quite difficult to control. Nitrogen can come from the nitrogenous waste in human and animal excrement as nitrites, nitrates, and ammonia. It can also come from the nitrogen in the atmosphere, which is converted into a usable form by the so-called "nitrogen fixation" organisms. Figure 1 also shows that these nitrogen fixation organisms need boron, calcium, and molybdenum. Carbon, the element required in the largest amount, can originate from the diffusion of carbon dioxide from the atmosphere into the water; it can come from decaying vegetative matter, including the dead algae which grew in the lake; it can also arise from rotting organic matter carried into the lake by land drainage. Other important sources are household sewage and agricultural and industrial wastes.

One of the largest sources of phosphates is sewage effluent; the other is land drainage and agricultural runoff. Most authorities agree that each source contributes about half of the phosphates in our waters. Since it is difficult to control phosphates from land drainage and agricultural runoff, let's consider the phosphates in sewage effluent. If most of the phosphate in sewage effluent came from detergents, its elimination from that source would solve the problem simply and easily. Unfortunately, depending on the locality, detergent phosphates contribute only about 28.5–70% to sewage effluent. The generally accepted figure is 66%, or about two-thirds of the total. The other third comes mostly from human waste. Every day we consume phosphates in our food and excrete a corresponding amount. This consumption and excretion also applies to cats, dogs, and other animals. Some phosphates also enter sewage systems through the waste disposal units in kitchen sinks.

The range of phosphate concentrations in sewage systems contributed by detergents, as reported in the literature, is based on various surveys and actual measurements in selected communities in this country. I checked these figures with some rough calculations to see whether my values were significantly different—that is, by an order of magnitude—from those I read. Physiology texts say that with an average American diet, each of us consumes and excretes approximately 1.4 lbs phosphorus per year. With a U.S. population of approximately 200 million, this amounts to about 280 million lbs of phosphorus per year. The total sodium tripolyphosphate manufactured per year in this country is approximately 2200 million lbs. Some is for export, and some goes into uses other than detergent. Let's assume that about 90%, or 2000 million lbs of sodium tripolyphosphate, goes into detergent. Since each mole of sodium tripolyphosphate contains 25.3% phosphorus, 2000 million lbs of it is equivalent to 505 million lbs of phosphorus. The sum total of phosphorus contributed by human excrement and detergents therefore is 280 plus 505 million lbs—*i.e.,* 785 million lbs of phosphorus. Dividing 505 million lbs into the total of 785 million lbs shows that the contribution of phosphorus from detergents is 64%. Based on these rough calculations, the figures reported in the literature are within reason.

How much phosphorus is needed for algal growth. Dr. Lange found that when other required nutrients are present in excess, he could grow blue-green algae in as little as seven parts of phosphorus per one billion parts of water (7 ppb). This figure agrees well with the findings of others. For example, Clair N. Sawyer, one of the most often-quoted authorities, reported that undesirable algal growth will occur at approximately 10 ppb of phosphorus. With these figures in mind, let's calculate the amount of phosphorus which is available to our streams and lakes. A task group on Nutrient Associated Problems in Water Quality and Treatment of the American Water Works Association, headed by Prof. P. L. McCarty of Stanford University, reported that the amount of phosphorus in the average effluent from a sewage treatment plant (which removed about 30% of the phosphorus) ranges from 3.5 to 9 mg/liter. This is equivalent to 3500–9000 ppb phosphorus. Assuming that 66% of this phosphorus is from detergents and is removed, about 1200–3100 ppb is left. Assume that this sewage effluent, on its way through streams to lakes, is diluted with water of lower phosphorus concentration. Assume also that some phosphorus is lost by reaction with minerals in the stream so that only 1% of the original phosphorus arrives at the lake. With these assumptions, there would still be 12–31 ppb phosphorus entering our lakes from sewage effluent which is of non-detergent origin. The calculations above support those sanitary engineers, researchers in environmental quality laboratories, civil engineers, zoologists, and biologists who state that even if all the phosphate were removed from detergents, there would still be more than several times the amount needed to support healthy algal growth from domestic sewage alone. More specifically, Dr. William J. Oswald, Professor of Public Health and Sanitary Engineering, University of California, Berkeley, said:

Independent of any detergent source, the average domestic sewage contains sufficient phosphorus from uncontrollable origin to support the growth of 1000 milligrams per liter of blue-green algae. Such an algal concentration is 50 times that ever found in Clear Lake, California, and 100 times that found in Lake Erie, and 1,000 times that found in the oceans.

The problem of reducing algal growth by limiting the phosphorus concentration in our waters is further complicated by the fact that other sources of phosphorus exist in addition to domestic sewage—*e.g.,* land drainage. Streams in three forested areas in Washington state had an average soluble phosphorus concentration of 0.007 mg per liter, or 7 ppb.

Phosphorus from agricultural run-off varies with locality. The minimum comes from normal land drainage; medium amounts occur where fertilizers are heavily used; and maximum concentrations are found in run-off from animal feedlots. Modern chicken farms are also great contributors of nutrients for algal growth. The Environmental Pollution Panel of the President's Science Advisory Committee (1965) showed that a cow generates as much waste as 164 humans, one hog as much as 1.9 people, and seven chickens present as great a disposal problem as one person. Altogether, U.S. farm animals produce 10 times as much waste matter as people.

To make the situation even worse, algae are extremely efficient users of phosphates. When they die, 30–70% are readily decayed by associated bacteria, releasing phosphates and other nutrients to support new growth. The remaining debris settles to the bottom and decays gradually over a period of years. A lake with a long history of algal growth can actually continue to support new growth even if no fresh nutrients, phosphate included, are introduced.

Some of the laws proposed or passed require a total ban on phosphates in detergents, and others limit phosphorus content in detergent to 8.7%. Since many non-chemists cannot differentiate phosphorus from phosphate, let's recall that the content of phosphorus, the element, in sodium tripolyphosphate (STPP), the phosphate normally used in detergents, is 25.3%. An 8.7% phosphorus content, therefore, is equivalent to 34% STPP. The average concentration of STPP in detergent is around 45%. The new laws thus propose a cut of STPP in detergent from around 45 to 34%, or a reduction of the phosphate contribution from detergents to our sewage systems from 66 to 50%, or a reduction of the phosphorus in our sewage to 84% of the original total. To expect that this reduction will help solve the algal growth problem is scientifically unsound. Fortunately the ban on phosphates does not include dishwashing compositions. Modern dishwashing machines just won't work efficiently without phosphates.

The above discussion assumes that phosphorus is the limiting element for algal growth. This is not always true for many bodies of water in the United States, or, in fact, in the world. Dr. Lange used Lake Erie water to study more than 1260 separate cultures of algae. He simulated the influx of each of the 16 essential elements by enriching each culture in the Lake Erie water with one of the elements. His results indicated that influx of nitrogen was growth-enhancing twice as often as that of phosphate, and that iron (as a soluble complex) and cobalt stimulated algal growth just as often as did phosphate.

Algal growth depends on photosynthesis, and photosynthesis does not occur without carbon dioxide. Thus, carbon could be the limiting element. If luxuriant algal growth depended on atmospheric CO_2 alone, this source of carbon would be the limiting factor. During a short period of vigorous growth, algae in the upper 12 inches of a lake can increase from 5 to 55 mg per liter. The amount of CO_2 needed has been calculated to be insufficient from the diffusion of atmospheric CO_2 into the water; diffusion would be too slow. The present theory is that a large portion of CO_2 is supplied by the bacterial decay of the dead algae and other organic matter. The symbiotic relationship between bacteria and algae is interesting. Algal growth by photosynthesis produces oxygen needed by the bacteria. Bacterial decay of organic matter produces CO_2 needed by algae.

Another source of carbon is the dissolved bicarbonates and carbonates in water. Many researchers on algal growth, including Pat C. Kerr and her co-workers, of the Federal Water Quality Administration of the U.S. Department of Interior, have shown that increase in the availability of inorganic carbon such as CO_2 and bicarbonate ions (HCO_3^-) rather than phosphate is responsible for algal growth in the waters they studied. For their studies, therefore, carbon is the limiting element.

Many entrepreneurs have introduced nonphosphate detergents. Most of these products use soda ash (sodium carbonate, Na_2CO_3) and sodium silicates to replace phosphate. Many lawmakers even propose a return to the use of sodium carbonate–soap combinations. When this extra supply of carbon from soap and the carbonates enters bodies of water where carbon is the limiting element, it will, of course, enhance algae growth.

Sodium tripolyphosphate, STPP, is an important builder in synthetic detergents. Among other functions it softens water by tying up hardness-causing metal cations into water-soluble complexes which don't interfere with cleaning and can be easily rinsed off. STPP also gives a washing solution with a pH of around 9–9.8, that is, alkaline enough for efficient dirt removal but not caustic enough to injure the skin. The result is extraclean clothing. The preference of American housewives for this new detergent over soap has practically eliminated laundry soap from the market. Soap is still very effective, however, in soft or softened water.

With the growing publicity on phosphate in detergents as a possible cause of eutrophication, detergent manufacturers first used the sodium salt of nitrilotriacetic acid (NTA) as a substitute builder. This compound is also a water softener. However, government reports suggested that when NTA is undecomposed in our soil or water, it can form complexes with mercury or cadmium. If these poisonous complexes entered our food, they might pose a health hazard. Many scientists agreed that the government report was based on a study which was not statistically significant. However, since doubt was cast on the safety of NTA, the Surgeon General of the Public Health Service of the United States suggested that the detergent industry withhold the use of NTA until more data are available.

Detergent pollution from municipal sewage treatment plant discharge in Pennypack Creek near Philadelphia, Pa.

Most of the other nonphosphate detergents contain soda ash (sodium carbonate, Na_2CO_3) and sodium silicates (sodium disilicate, $Na_2Si_2O_5$; sodium metasilicate, Na_2SiO_3) as builders. They soften water by removing the "hardness"-causing cations as insoluble precipitates, *e.g.*, calcium carbonate ($CaCO_3$), magnesium silicate ($MgSiO_3$), and calcium metasilicate ($CaSiO_3$). Not all of these precipitates are rinsed off; some are trapped in the fabric. These trapped salts interfere with the flame-proofing property of the phosphorus-containing flame retardants. Presently available flame retardants for children's pajamas all contain phosphorus compounds. The theory is that upon pyrolysis, the phosphorus compounds decompose into phosphoric acids, which in turn catalyze the decomposition of the cellulose fabric into very slow-burning chars. The trapped salts on the cloth from nonphosphate detergent convert the phosphoric acids into phosphate salts which are not effective as catalysts for converting burning cellulose to char. Thus, the pajamas burn.

How well do the non-phosphate detergents clean? A limited test was done by a consumer organization using about 300 families. They were given unlabeled phosphate and nonphosphate detergents. After about a month, most of the housewives using the nonphosphate detergents were about as satisfied as those using the phosphate detergents.

In the laboratory, however, cleanliness can be measured more precisely under carefully controlled conditions. Artificially soiled cloths are washed with various phosphate and nonphosphate detergents, and cleanliness or whiteness is measured by a reflectometer. Such data are reproducible. Results from many laboratories show that phosphate-built detergents in hard water indeed clean better. Friends of mine who travel extensively in the Far East where phosphate detergents are not prevalent, did not notice much change in their shirts laundered abroad until they returned and found that the shirts left at home were much whiter. However, this is not scientific proof that phosphate detergents give whiter shirts.

From the health hazard point of view, commercially available phosphate detergents are less caustic or less basic than those built with soda ash and sodium silicates. The pH of laundry water solutions which contain phosphate detergents is around 9 to 9.8 while those with soda ash and sodium silicate are around 10.5–11.5. Since the pH scale is logarithmic, an increase of one unit, *e.g.*, from 10 to 11, means an increase of 10 times in basicity. Highly caustic or basic material is injurious to the sensitive skin of our eyes, nose, and throat.

In September 1971, Dr. Jesse L. Steinfeld, then the Surgeon General of the Public Health Service, told Congress that the federal government might eventually have to ban detergents built with soda ash because of their health hazard. He also advised housewives to use phosphate detergents because they are now the safest. Dr. Steinfeld's statement was supported by Russell E. Train, Chairman of the President's Council on Environmental Quality, Dr. Charles C. Edwards, the head of the Food and Drug Administration, and William D. Ruckelshaus, administrator of the Environmental Protection Agency. They strongly urged state and local authorities to reconsider laws and policies which unduly restrict phosphates in detergents.

Eutrophication of streams and lakes cannot be solved by simply banning phosphates. Temporarily expedient solutions may cause more harm than good. As Secretary of Commerce Maurice H. Stans said succinctly in his address to the National Petroleum Council on July 15, 1971:

. . . Laws to ban phosphate detergents may give the public the notion that the problem is solved, while nutrients, including phosphates, continue to flow into the lakes and rivers from other sources—agricultural and natural as well as manmade. And some of these cannot be controlled. So if people assume that just a legal ban on phosphate detergents will do the job, they may only lull themselves into neglecting far more significant scientific efforts to help purify our waters through phosphate removal techniques in municipal waste treatment plants.

Do we now possess the technology to remove phosphates and the other major nutrients for algal growth in sewage treatment plants? The answer is yes. Are we using this technology? Very little. Sewage treatment practices, though they vary with locality, consist in general of three separate steps. The primary step is the removal of large solid particles from the raw sewage by settling or by filtration through coarse screens. The effluent from this step, which still contains the fine solids, organic waste products, and dissolved minerals, is passed into the secondary treatment step. Here the organic matter is decomposed by air and bacterial action. The secondary treatment step employs one of two general methods. The first, a trickling filter process, brings air and the liquid waste together and passes the mixture through a pebble and sand filter

Aerial view of the $16 million biological wastewater treatment facilities of the Rahway Valley Sewerage Authority—one of the largest secondary wastewater treatment complexes built in New Jersey in the last decade. At left, foreground, two new aeration tanks, larger than a football field. Right, foreground, four new 120-ft diameter final settling tanks. Center, the new Pump and Blower Building, housing the principal control center for the complex.

bed loaded with attached microorganisms. Much of the organic waste is thus decomposed by bacterial action and air oxidation. The second step, the activated sludge process, destroys organic waste by bubbling air through the sewage liquor containing a suspension of microorganisms. The sludge thus formed settles and is discarded as waste.

When carried out properly, the secondary treatment step removes most of the soluble organic matter as volatile carbon dioxide gas; thus, if the treated liquor is discharged into streams and lakes, it is not a source of organic carbon to support algal growth. It also does not decay further, using up dissolved oxygen in the water. In other words, it would have a low BOD (biological oxygen demand). The secondary treatment step also removes as much as 20–30% of the phosphates as insoluble precipitates (mostly as calcium and magnesium phosphates).

The treated liquor from the secondary treatment step still contains much phosphate and other dissolved minerals. It can now go to another plant for the tertiary treatment step if such a plant is available. Phosphate can be removed here as the insoluble tricalcium phosphate or hydroxyapatite by the addition of lime, or as the insoluble iron phosphates by the addition of water-soluble ferrous or ferric salts, or as the insoluble aluminum phosphates by the addition of a soluble aluminum salt. Such treatments, coupled with filtration through a bed of carbon in which the carbon has been specially treated or activated for absorbing impurities, can remove up to 98% of the phosphates. Such treated water, after chlorination to kill the residual bacteria, is suitable for drinking. A sewage treatment system which uses modern technology with good results is that of Lake Tahoe, on the California–Nevada border.

Unfortunately, many municipalities have no sewage treatment plants, and their waste is discharged directly into the nearest stream, river, or lake. In some cities, only primary and secondary treatment are available, and some cities have only primary treatment. These sewage treatment facilities cannot keep pace with the population growth and are overloaded. Consequently, much of their effluent is only partially treated.

In many areas domestic sewage goes into individual septic tanks, where organic waste is decomposed by bacteria, and the effluent seeps into the ground. Here, most of the phosphate reacts with the soil and precipitates out as insoluble salts. However, in some cases bacterial decomposition of the organic waste is incomplete. For example in Suffolk County, Long Island, New York, where more than a million people live, most of the household sewage goes into individual septic tanks. Sometimes, because of incomplete decomposition, some of the sewage water carrying residual organic matter percolates through the sandy soil and enters the water supply. If some of the organic waste is from the surfactants in the detergents, water drawn from a faucet will

Little Quinton (Mogey) DeMann of Wilmington, Del., pensively ponders hooking one of those beautiful rainbow trout swimming in what was once raw sewage. The water was made sparkling clear by a unique new municipal treatment plant being built inside a neighborhood house by AWT Systems, Inc. and ITT Levitt.

be foamy. Thus, a recent law forbids the use of synthetic detergents altogether in Suffolk County. Unfortunately, this law removes only the indicator showing that contamination from sewage has occurred but not the contamination itself.

Some localities are reluctant to build sewage treatment plants because they are expensive. Depending on the percentage of phosphate one wants to remove from sewage effluents, the cost of the many systems proposed varies from 2.5¢ per 1000 gallons for 90% phosphate removal to 3.7¢ per 1000 gallons for 98% phosphate removal. Mr. Ruckelshaus stated that based on a survey of 12,500 sewage plants, it would probably cost from $1.70 to $3.35 a year for each person living near the sewage plant, to operate the phosphate treatment system. Estimates for building sufficient sewage treatment plants in this country run as high as tens of billions of dollars. In fact, the estimate by EPA in 1973 put the total at $60 billion by 1990.

Many laboratories are now working to develop more efficient and economical methods for treating sewage waters. Phosphate removal is only part of their program since phosphate is only one of the many nutrients in sewage waste which supports algal growth. Better and cheaper methods for cleaning our sewage certainly is a challenge to chemists, biologists, and engineers.

The development of an efficient, safe, and economical builder to replace phosphate in detergent is also a challenge. With such a builder available, provided that it does not present a removal problem, the size of the equipment and the amount of chemicals needed for phosphate removal in a sewage treatment plant would be reduced.

Can agricultural runoff be controlled so that we can return to the environmental standards of our forefathers? I think such controls are possible. For example, large basins might be built to collect runoff from areas of intensive cultivation for treatment before allowing it to flow into the neighboring streams and lakes. Similar collection basins might also be built to collect runoff from animal feedlots.

The problem of controlling waste effluents from animal feedlots is a large order of magnitude greater than that from agriculture runoff and is probably more readily accomplished. Even so, the control of animal feedlot runoff is not a minor task. It has been estimated that on a rainy day a 50-acre feedlot can produce runoff equivalent to the raw sewage produced by a city of 60,000 people; and annually more than a billion tons of manure are produced by the animal feedlots. Anti-pollution laws are gradually forcing livestock men to devise ways to collect and use animal waste and prevent it from contaminating the environment. To comply with those laws, some feedlot operators have resorted to housing the cattle in huge metal barns with pens bedded with a few inches of raw sawdust or ground bark. Periodically the manure and bedding mixture is collected and, through bacteria breakdown, converted into pasteurized and deodorized garden fertilizer. Other methods under development include collection of the cattle manure and leaching out of its protein content. The residue is then fermented into roughage. It is claimed that the combined protein and roughage obtained are comparable in flavor and food value with corn silage. Poultry manure has been similarly processed into cattle feed. All these indoor manure treatment plants represent billions of dollars nationwide. Eventually, we, the consumers of agricultural commodities, will have to pay for it by paying much more for our food.

Eutrophication is a complex process. We do have the technology to build efficient sewage treatment plants to remove phosphates and other nutrients which support algal growth. We should continue our research to develop improved technology for an improved environment.

9

Metal Treating and Cleaning

Phosphatizing Metals

Modern cars, refrigerators, washing machines, and other electrical appliances with painted or enameled surfaces all wear phosphatized undercoatings to prevent the paint from blistering and peeling.

Phosphoric Acid. In phosphatization, a metal surface (iron, steel, zinc, or aluminum) is coated with a solution of phosphoric acid containing a metal (zinc) salt. This smooth coating retards corrosion and provides an excellent base for paint.

A phosphoric acid solution containing manganese salts is used to coat objects that will not be painted. This coating is thick and absorbent and provides excellent corrosion resistance. To increase protection, a thin film of oil or wax is usually applied to the absorbent surface. This black, slightly oily surface is seen on nuts, bolts, screws, tools and machine parts.

Manganese phosphatized coatings are also used extensively for automobile engine parts such as pistons, piston rings, tappets, cams, and gears. The film, wetted with machine oil, acts as a wear-resistant coating on surfaces subject to constant friction. The wear then produces a very smooth burnished surface, and scoring is reduced to a minimum. In an automobile engine, the phosphatized coating may be completely worn off during the break-in period, but it is replaced by phosphorus containing materials, usually zinc dialkyl phosphorodithioates added in motor oils (Chapter 14).

Phosphatizing is also used to coat sheet metal that is subsequently drawn, stamped, or cold worked (*e.g.*, bottle caps). Here, the microcrystalline phosphated coatings act as lubricants during the press operations and keep the metal from cracking.

Modern phosphatizing is a result of technologies developed over the last 100 years. As early as 1869 a British patent was issued to prevent corset stays (thin metal strips) from rusting because of damp air or perspiration. The stays were plunged red hot into a phosphoric acid solution, but the resulting coating was thin and not very durable. In 1907, T. W. Coslett patented the first improvement on this process. He dissolved iron filings in dilute phosphoric acid; this coating consisted of a mixture of corrosion-resistant iron phosphates. Other modifications were later introduced. One significant improvement was the replacement of iron filings with zinc or manganese salts. A zinc salt produced a smooth coating of fine zinc–iron phosphate crystals. A manganese salt produced a coating chiefly of relatively coarse (absorbent) crystals of manganese phosphate along with iron phosphate.

Coatings are formed by the interaction of phosphoric acid and iron at the metal surface. The primary zinc–iron phosphates and manganese–iron acid phosphates are soluble in the highly acidic phosphatizing solutions. In weak acid the soluble primary phosphates are converted to the less soluble tertiary zinc phosphate [$Zn_3(PO_4)_2$] and secondary iron phosphate ($FeHPO_4$) and precipitate out. For example, when an iron object to be coated is dipped in a phosphoric acid solution containing zinc or manganese salts, the surface of the object reacts with phosphoric acid and consumes some of the phosphoric acid next to the surface of the object. Since the acidity next to the surface is now lowered, the tertiary and secondary zinc-iron phosphates or manganese-iron phosphates, which are less soluble, precipitate to form the coating.

The major drawback of this process is that it takes too long—usually more than two hours at 180°–200°F (82°–93°C). When a solution is

When this housewife went shopping for a new refrigerator, she looked for a product with good appearance, trouble-free operation, and convenience. The fact that this refrigerator wears a phosphatized undercoating to keep its paint from blistering and peeling didn't particularly concern her. However, modern technology has made it easier for her to find exactly what she was looking for.

Courtesy General Motors Research Laboratories

Zinc phosphate coating on steel

Individual crystals of zinc phosphate are nucleated from sites on the metal surface and grow outward and laterally. A density is achieved that provides a large number of interlocking anchor-like surfaces for good paint adherence and corrosion resistance. The phosphated surface also has a capillary effect which enhances wetting and promotes uniform coating.

Underside of electrodeposited paint primer

An improved approach to studying the interaction of paint and a zinc phosphated surface is to dissolve the phosphate coating and free the paint film. This allows direct examination of the underside paint film surface which had been in contact with the phosphated surface. The mechanical interlocking of the "cast" primer resin surface revealed by this technique illustrates the mechanism by which paint adherence is improved and surface area of bonding increased. Replication of the fine veining present on individual crystals confirms the excellent wettability provided by the surface. With the aid of such capability, the paint engineer has been able to incorporate improved adhesional and cohesional properties into each of the layers so the finish performs as a unit rather than as individual layers.

freshly prepared, complete coating only takes about 10–15 minutes. But as the bath is used, coating of each successive metal piece takes longer. The slowdown results from accumulation of ferrous ions (Fe^{2+}) in the bath. Some of the iron piece being phosphatized is dissolved by phosphoric acid, and the more iron pieces that are phosphatized, the more ferrous ions accumulate. The net effect is a gradual change in the composition of the acid solution from its original zinc– or manganese–phosphate content to one with zinc–ferrous phosphate or manganese–ferrous phosphate.

Experiments showed that ferrous phosphate in phosphoric acid retards coating formation. A pure zinc phosphate–phosphoric acid solution coats fastest; next is the manganese phosphate–phosphoric acid solution. The obvious answer is to remove ferrous ions from the phosphatizing solution, but phosphatizing an iron surface inevitably means that some iron will dissolve to give ferrous phosphate.

Fortunately, *ferric* (Fe^{3+}) phosphate is insoluble in the phosphatizing bath. The method developed to remove ferrous ions, therefore, involves converting them to insoluble ferric phosphate by adding a small amount of an oxidizing agent (sodium nitrate). Thus, the coating time is reduced to less than five minutes, depending on operating temperature and other solution characteristics.

Accelerators have also been developed. Addition of 0.002–0.4% copper as a soluble salt such as copper nitrate accelerates the rate several fold. One theory behind this action is that the copper catalyst plates out in minute specks as metallic copper on the metal surface to be coated. This sets up a multitude of small batteries causing the surrounding metal to dissolve at an accelerated rate and thus react more rapidly with the solution to form the coating. Nickel and cobalt ions also accelerate coating formation, but the mechanism of their reaction is not well understood. Other modifications for phosphatizing solutions are described in "Protective Coatings for Metals," by R. M. Burns and W. W. Bradley (Reinhold, New York, 1967).

Phosphatized coatings can be applied to surfaces by dipping the objects into the solution—*e.g.*, the manganese–phosphate solution coating of nuts and bolts. The coating weighs about 2.5 grams per square foot and is about 0.0002 inch thick. Phosphatized coatings can also be sprayed on large objects such as automobiles, refrigerators, and washing machines. Zinc phosphatization gives a light gray, smooth, slate-like textured surface that weighs about 0.2 gram per square foot and is about 0.000025 inch thick.

Courtesy Ford Motor Co.

This automobile body is being electrocoated with rust-inhibiting primer containing phosphates to give its metal surfaces a thorough, uniform coating

Monosodium Phosphate. To phosphatize a metal surface with less rigid final requirements, a formulation containing monosodium phosphate is used. Monosodium phosphate is prepared by the reaction between sodium carbonate and a strong solution of phosphoric acid in water:

$$2 \; H_3PO_4 + Na_2CO_3 \longrightarrow 2 \; NaH_2PO_4 + H_2O + CO_2$$

Since only one hydrogen atom in the phosphoric acid is replaced with a sodium atom, the remaining product with two hydrogens is still an acid. The major applications for this compound are based on the fact that it is a solid acid that dissolves readily in water. For iron surfaces, monosodium phosphate is generally used in combination with a detergent; the detergent is absorbed into solid monosodium phosphate particles. To do this, the monosodium phosphate must be produced in a porous form by spray-drying or drum-drying its water solution. Monosodium phosphate obtained by crystallization from a solution would not be porous and therefore not suitable for this application.

For many indoor applications—*e.g.*, office furniture—where weathering is not a major problem, a high quality protective undercoating such as that from zinc–phosphoric acid phosphatization is not necessary. Only a treatment that cleans the metal surface and makes it chemically acceptable to paint is needed. Thus, the metal surface is sprayed with an aqueous solution containing about two ounces per gallon of a solid mixture of monosodium phosphate, sodium acid pyrophosphate, and some detergent. This solution is acidic (pH 3.8–4.1) and reacts well with the iron surface. The iron phosphate coating formed is extremely thin (about 50 mg per square foot), but it is enough to make the surface cling to the paint. The detergent in the formulation cleans any dirt spots.

Metal Cleaning

The phosphoric acid–organic solvent cleaning process is a mild form of phosphatizing. This process uses a mixture of organic solvents such as alcohol or butyl cellosolve ($C_4H_9OCH_2CH_2OH$, an ether alcohol) with phosphoric acid plus a wetting agent. The treatment removes greasy contaminants and destroys light rust and annealing scale on a metal surface. It also gives a mild phosphatic coating to the surface which promotes paint adhesion.

A little known, but ingenious use of dilute phosphoric acid is to remove rust safely from chrome strips on older cars. The most readily available source of this solution is the cola drinks. As described in chapter 4, a cola drink contains 0.057–0.084% of 75% phosphoric acid. It is quite a sight to see a man on a hot summer day take a drink of cola and then pour some of the same drink on a rag and use it to polish the rusty chrome parts of his car.

Cleaning Boilers

Phosphoric acid has recently found a new use in cleaning power-plant boilers. In modern boilers it removes rust and mill scales; in old ones it removes mineral water deposits such as calcium and magnesium carbonates. Formerly, inhibited hydrochloric acid (HCl) was used. (An inhibited acid is formed by the addition of an organic amine, which retards acid attack on the metal surface.)

Hydrochloric acid presents disadvantages. If the temperature is not kept below 150°–160°F (66°–71°C), the amine inhibitor does not work. Since HCl is quite volatile, its vaporized portion is not inhibited, and it can attack the metal. This volatility also produces a corrosive acid atmosphere in the power plant that is unpleasant for the workers and endangers delicate and expensive control instruments. Finally, a surface cleaned by hydrochloric acid tends to rust rapidly when exposed to a humid atmosphere.

Phosphoric acid is an effective boiler cleaner and does not have these disadvantages. A 5% solution can be boiled in a power-plant boiler at atmospheric pressure to remove deposits. The cleaned surface which now wears a thin coating of iron phosphate then resists rusting. Statistics on new boiler cleaning are impressive. A boiler commonly used by electric power plants has a normal steaming capacity of about 600,000 lbs per hour and a steam pressure of 900 psi at 900°F. For a boiler with such a capacity, 14,850 lbs of inhibited 75% phosphoric acid are required. Before acid treatment, the boiler must first be heated with an alkaline solution containing 500 lbs trisodium phosphate dodecahydrate [(Na$_3$PO$_4$-12 H$_2$O)$_5$NaOH] along with sodium sulfite. After acid cleaning, it is rinsed with 0.1% phosphoric acid and then conditioned with a solution of tetrapotassium pyrophosphate (K$_4$P$_2$O$_7$) and potassium hydroxide (KOH) along with some potassium sulfite. The acid boil-out takes eight hours, and the drained solution contains one part per thousand of iron, equivalent to more than a ton of iron oxide (Fe$_2$O$_3$).

Mirror-Like Finishes on Metals

Chemical Polishing of Aluminum. Polishing gives metal surfaces a bright, shiny, mirror-like finish. Since mechanical polishing requires considerable labor, it is expensive as well as impractical for irregularly-shaped articles. Consider, however, chemical polishing, or "bright dip." This process is especially suited to objects made of aluminum or its

Courtesy General Motors Co.

Chemical polishing or "bright dip" is especially suited to objects of aluminum or its alloys, for example the shiny trim on this automobile

alloys—*e.g.*, automobile trims. These trims have now almost completely replaced chrome-plated trims. Other bright dip objects are some automobile grills, light reflectors, and handles for refrigerators, washers, and freezers.

In the bright dip process the article is immersed in a hot solution (91°–99°C) that is about 95 parts 85% phosphoric acid and about four to five parts of 68% nitric acid (HNO_3), along with 0.01 to 0.04% copper nitrate. It is said that in some plants, the operators simply throw in a handful of pennies though the legality of this practice is questionable. Sometimes they also add a trace of nickel and a little wetting agent. The bright dip process now takes less than three minutes.

Additives are important in polishing high purity aluminum pieces. Without them, the process tends to emphasize the grain of the metal, and the surface is not as shiny as possible. The wetting agent brings the polishing solution and the metal surface into closer contact and increases the reaction rate, thus improving metallic luster. Since the bright-dip bath is thick and sticky, each piece drags out some solution after it is dipped. Added wetting agent lowers the surface tension of the solution and thus reduces drag-out. Although nitric acid in the bath evolves toxic brown fumes of nitric oxide during operation, additives such as ammonium compounds eliminate most of them.

The composition of the bright-dip bath determines the quality of the finish. One important factor is aluminum content, and its optimum concentration is 25–30 g/liter. Since chemical polishing of aluminum means that some of the metal dissolves, the aluminum content of the bath increases with usage. The higher the aluminum content, the higher must be the concentration of phosphoric acid for optimum brightness. However, an acid concentration which is too high will pit the surface. If the concentration is too low, the surface tends to be dull. The effect of dissolved aluminum content at various acid concentrations (as indicated by the density of the solution) on the brightness of the article, is shown in Figure 1.

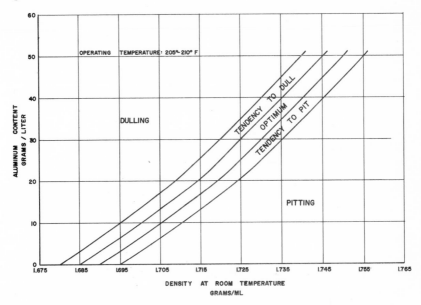

Figure 1. Effect of dissolved aluminum content on brightness of article at various acid concentrations (shown by density of the solution) in a bright-dip bath

The operator of the bright-dip bath must measure the concentrations of aluminum, phosphoric acid, water, and nitric acid and adjust them to keep the bath operating at optimum conditions. Chemists are still looking for additives which will permit them to operate this type of bath at a wider range of aluminum or acid concentrations.

What actually happens during bright-dipping? Some metallurgical chemists believe the action is based on nitric acid oxidation of the aluminum surface to give a porous aluminum oxide film that is then dissolved by phosphoric acid. In this reaction, the microscopic mountains on the surface dissolve faster than the valleys. The process thus converts a relatively dull surface to an ultrasmooth, mirror-like finish.

In commercial practice, the bright-dipped article, after rinsing, is generally anodized in an electrolytic bath to give the bright surface a protective coating of almost transparent aluminum oxide.

Aluminum surface highly magnified, showing mountains and valleys.

Aluminum surface oxidized by nitric acid to form nodules of aluminum oxide (Al_2O_3). The mountains are attacked more rapidly than the valleys.

$$Al + 2\ HNO_3 \longrightarrow \tfrac{1}{2}Al_2O_3 + 2\ NO + H_2O$$

Aluminum surface after dissolution of aluminum oxide by phosphoric acid.

$$Al_2O_3 + H_3PO_4 \longrightarrow 2\ Al(H_2PO_4)_3 + 3\ H_2O$$

Electropolishing of Metals in Phosphoric Acid Solution. Like mechanical polishing and chemical polishing, electrolytic action can also polish metal surfaces. This third method—electropolishing—is especially suited to stainless steel articles. Shapes can be regular or irregular, such as forks, knives, spoons, surgical instruments, and refrigerator shelves. The process is done in an electrolytic bath containing either a solution of 75% phosphoric acid or a mixture of phosphoric acid with a small amount of sulfuric acid. A mixture of phosphoric, sulfuric, and glycolic acids is used in one commercial solution. The article is suspended from the anode of this bath, and an electric current is passed. The temperature is kept between 160° and 180°F (71°–82°C), and the time varies from 1 to 10 minutes, depending on the roughness of the surface. The microscopic mountains on the surface are leveled by the combined action of the acid and electric current to give the desired finish.

Electropolishing can also be used for aluminum. However, since the cost of the equipment is much higher than that for chemical polishing, the latter is used more often.

10

Flame Retardants

Halloween brings the familiar sight of children dressed as goblins, monsters and ghosts in brightly colored costumes. Since the children sometimes carry candles and jack-o-lanterns, one might wonder why their paper costumes don't catch fire.

Paper dresses for everyday wear have also been a recent fad. Since the women who wear them often smoke, one might also wonder why no accidental fires have started from these dresses. The answer is that halloween costumes and paper dresses share a common characteristic— they are difficult to burn. The agents responsible for this property are the ammonium phosphates. Also used as fertilizers and in dyeing textiles, they are today's most widely used flame retardants.

Ammonium Phosphates

Ammonium orthophosphates are prepared in ways similar to sodium and potassium orthophosphates (discussed earlier). The hydrogen ion in orthophosphoric acid is replaced with a monovalent cation—in this case, the ammonium ion. In many other respects, however, the preparation is quite different. Ammonium hydroxide is a much weaker base than sodium or potassium hydroxide and forms an unstable salt with phosphoric acid. When ammonium salts are heated, ammonia gas is released. Hence, monoammonium and diammonium phosphate are the only ammonium orthophosphates stable enough to be commercially important.

Triammonium phosphate can be prepared at low temperatures, but at room temperature it reverts rapidly to ammonia gas and diammonium phosphate. Diammonium phosphate, when heated to about 70°C, decomposes to ammonia and monoammonium phosphate. Although the latter is the more stable of the two, it also decomposes to ammonia and phosphoric acid when heated. For example, at 125°C a measurable pressure of ammonia is evolved (0.05 mm Hg). This thermal decomposition of ammonium phosphates is not always undesirable. As shown later, many applications for these compounds depend on it.

Commercial quantities of monoammonium phosphate are made by pumping dried ammonia gas into an 80% solution of phosphoric acid. Product composition is controlled by maintaining the pH at 3.8 to 4.5. (If the pH is too high because of excess ammonia, some diammonium phosphate forms.) On cooling the mixture, crystals of monoammonium phosphate precipitate out. They are separated by centrifugation, dried, and the mother liquor is recycled for the next charge.

Diammonium phosphate is prepared by bubbling two moles of ammonia gas into one mole of 80% phosphoric acid solution. The second mole of ammonia is added at temperatures below 50°C to prevent thermal decomposition; pH is controlled to about 8. Crystals formed on cooling are centrifuged and then dried at below 50°C.

Tetraammonium pyrophosphate can also be synthesized but not by direct thermal conversion of diammonium phosphate (as with tetrasodium and tetrapotassium pyrophosphates). When diammonium phosphate is heated to a high temperature, it dissociates into ammonia gas and monoammonium phosphate; further heating produces more ammonia gas and phosphoric acid. Tetraammonium pyrophosphate can be made by neutralizing crystalline pyrophosphoric acid with ammonium hydroxide.

Ammonium polyphosphate is also difficult to prepare by the high temperature conversion of monoammonium phosphate. However, it can

$$O$$
$$\|$$

be prepared by heating phosphoric acid with urea (H_2NCNH_2) under an ammonia atmosphere. It can also be prepared by neutralizing polyphosphoric acid with ammonium hydroxide. Ammonium polyphosphate, unlike the ammonium orthophosphates or pyrophosphate, is practically insoluble in water. This water insolubility makes it useful as a flame retardant in paint coatings and in non-woven fabrics where it is bonded permanently by polymeric binders.

Since monoammonium phosphate and diammonium phosphate can be used interchangeably in most of their applications, they are discussed together here. To understand how ammonium phosphates work as flameproofing agents, let us consider some theories on reducing flammability. Most of these theories are based on what happens when cellulose materials burn [cotton, paper, viscose rayon (regenerated celluose) and wood]. First, most of the material is converted to liquid fragments; further heating converts these fragments into volatiles which then burn. Only a small amount of the original cellulose decomposes into solid fragments which, upon further heating, decompose mostly to a slow-burning char (carbon).

In the presence of an acid the process is somewhat different. Most of the cellulosic material is first converted into solid fragments. Further heating changes these fragments to a slow-burning char. An acid seems to catalyze the thermal dehydration of the cellulose molecule—*i.e.*, long chains of repeating glucose units ($C_6H_{12}O_6$)—into carbon and water. This char or carbonaceous residue insulates the remaining portion from

Memphis, Tenn. officials watch cotton mattress flame retardancy test. Many people lose their lives each year in fires which are triggered by careless smoking in bed.

further burning. Therefore, only a small portion of the cellulose is converted to flammable gases.

Acids are the best agents for catalyzing the thermal degradation of cellulose in the direction of slow-burning char and less flammable gases. However, they disintegrate cloth or paper if applied directly. To circumvent this problem, the cloth or paper is treated with a non-acidic chemical which when heated decomposes into an acid. Mono- and diammonium phosphates decompose on heating to phosphoric acids. These acids then catalyze the decomposition of cellulose in the desired manner and extinguish the fire.

Another salt which generates nonvolatile acids is ammonium sulfamate. When heated, it liberates ammonia and sulfamic acid.

When antimony oxide is combined with certain chlorine compounds, such as chlorinated wax, and heated, it forms the acidic antimony trichloride, an effective flame retardant. Antimony trichloride catalyzes burning cellulose to form as great a percentage of the slower-burning char or carbon and less of flammable gas as the other salts mentioned above, but it is not an entirely satisfactory flame retardant. The char continues to burn (afterglow), and it is possible to get just as nasty a burn from a glowing char as from a flame. However, phosphoric acid-generating flame retardants are also afterglow retardants; the char formed does not continue to burn.

One theory advanced to explain this phenomenon is that the phosphoric acid flame retardant catalyzes a burning process that favors the formation of carbon monoxide rather than carbon dioxide. Formation of carbon monoxide gives off only 24.4 kilocalories of heat per mole while the formation of carbon dioxide gives off 96 kilocalories of heat per mole. The greater heat liberated with CO_2 formation supports further combustion and, in this case, afterglow burning. [A calorie is the amount of heat needed to raise one gram of water one degree centigrade. A kilocalorie is 1000 calories. Incidentally, a dietician's calorie is actually a chemist's kilocalorie. So an average size man needs a daily intake of food which when burned gives off 3200 kilocalories of heat.]

In flameproofing cloth and paper, ammonium phosphate can be applied by spraying as a water solution or by dipping. A 3–5% dry weight gain is generally needed for effective flame retardancy. However, since ammonium phosphate is water soluble, clothing loses its flameproofing after laundering. This problem is not serious for draperies and theatrical scenery, which are not laundered often and which can be re-treated after each cleaning. Neither is it a problem for disposable paper dresses and Halloween costumes. Thus, ammonium phosphates are used extensively to flameproof these articles as well as building materials, such as wood and cellulose insulations, and the interior of wallboard. This is the reason old lumber from demolished buildings often makes poor fuel.

As indicated in Chapter 2, paper book matches and wooden matches are all treated with ammonium phosphates to prevent afterglow. Christmas trees and decorations are occasionally treated to reduce fire hazards.

Many of the chemical formulations used as fire extinguishers contain ammonium phosphates. They usually also contain a thickener such as sodium carboxymethylcellulose which forms a gel when water is added. In fighting forest fires these flame-retardant gels can then be dropped from airplanes or helicopters in globs over the target area. They retard combustion by coating trees in the path of an advancing fire.

Urea Phosphate

Another industrially available flame retardant is urea phosphate. It is prepared by the reaction of one mole of urea ($H_2N\overset{\overset{\displaystyle O}{\|}}{C}NH_2$) with one mole of phosphoric acid:

$$H_3PO_4 \;+\; H_2N\overset{\overset{\displaystyle O}{\|}}{C}NH_2 \;\longrightarrow\; H_3PO_4 \cdot H_2N\overset{\overset{\displaystyle O}{\|}}{C}NH_2$$

phosphoric acid   urea

To flameproof cotton fabrics, the material is soaked in a solution of urea phosphate so that upon drying, the add-on, or gain in weight, is about 15% above the dry weight of the fabric. The dried treated fabric is then cured at around 160°C. Heat causes a chemical reaction between the phosphate and the cellulose molecules in the fabric. Since the phosphate is now chemically bonded to the fabric, it does not wash off. This treatment, however, partially degrades the fabric, causing it to lose about 20–50% of its tensile strength and even more tear strength.

Although the phosphate is bound more or less permanently to the fabric, after repeated washings the ammonium ion (from the decomposition of the urea) attached to the phosphorus atoms exchanges with the sodium ions from detergents and calcium ions from water. When enough ammonium ions have been replaced, sodium and calcium phosphates are left attached to the fabric, and flame resistance is lost since neither sodium nor calcium phosphate becomes acidic on heating.

Another interesting flame retardant is obtained by the high temperature reaction of P_2O_5 with gaseous ammonia. The resulting compound has a complicated and uncertain structure and hence no chemical name. In one form it is known by the trade name Victamide. It contains phosphorus–nitrogen bonds and $PONH_4$ groups. It dissolves very slowly in water, first forming a gel. Since it is less soluble than ammonium phosphate in water, it does not wash off as quickly in water. It is used to some extent in place of ammonium phosphates to flameproof paper, fabrics, and backings for plastic sheets.

Phosphines

Phosphine, PH_3, is such a poisonous gas that 2.8 mg in a liter of air is lethal within a few minutes. When prepared by most common methods, it contains an impurity ($H_2P–PH_2$) that makes it ignite spontaneously in air. A compound with these properties hardly seems a likely candidate for use in wearing apparel, yet one of the commercially available processes for flameproofing cotton is based on a compound prepared from phosphine.

Phosphines, which are compounds of phosphorus and hydrogen, are also known as phosphorus hydrides or hydrogen phosphides. Of the many members of this family, the most well-known is the first, PH_3, usually called phosphine. The others are diphosphines, triphosphines, and tetraphosphines.

$$PH_3 \qquad H_2P{-}PH_2 \qquad H_2P{-}\overset{\displaystyle H}{\underset{|}{P}}{-}PH_2 \qquad H_2P{-}\overset{\displaystyle H}{\underset{|}{P}}{-}\overset{\displaystyle H}{\underset{|}{P}}{-}PH_2$$

phosphine   diphosphine   triphosphine   tetraphosphine

Simple phosphine is fairly easy to prepare. However, as the number of phosphorus atoms in the compounds increases, preparation and isolation become increasingly difficult.

Some people believe phosphine occurs in nature. They have speculated that "will-o'-the-wisp," the flickering light seen over marshland at night, may result from the spontaneous ignition of marsh gas (methane, CH_4) by the presence of trace amounts of phosphines. Both marsh gas and phosphines evolve from decaying vegetable matter.

Chemists are not yet able to prepare phosphine in a practical way by combining phosphorus and hydrogen directly. The most common preparative methods are (1) reaction of certain metal phosphides with water, and (2) hydrolysis of phosphorus in a base such as sodium hydroxide or calcium hydroxide, $Ca(OH)_2$ (hydrated lime).

Metal phosphides are compounds of a metal with phosphorus. Aluminum phosphide (AlP) is made by igniting an equimolar quantity of powdered aluminum and red phosphorus. Zinc phosphide (Zn_3P_2) is made similarly by combining zinc and phosphorus directly. Calcium phosphide (Ca_3P_2) is synthesized by reducing calcium phosphate with aluminum powder at high temperature. The products are calcium phosphide (Ca_3P_2) and aluminum oxide (Al_2O_3). These three metal phosphides will liberate phosphines when water is added. For example:

$$2 \text{ AlP} + 3 \text{ H}_2\text{O} \longrightarrow 2 \text{ PH}_3 + \text{Al}_2\text{O}_3$$

Aluminum oxide is the by-product; the volatile gas burns spontaneously in air. Pure phosphine, PH_3, does not ignite spontaneously; ignition is caused by a small amount of the impurity diphosphine ($H_2P–PH_2$). This spontaneous-ignition property of impure phosphine gases is used in sea flares based on calcium phosphide.

Phosphine preparation by the hydrolysis of phosphorus in the presence of a base such as sodium hydroxide or calcium hydroxide is a complex reaction. Other products such as hydrogen (H_2), hypophosphite,

$$\left[\begin{array}{c} O \quad H \\ \| \quad \diagup \\ -O P \\ \diagdown \\ H \end{array} \right]^{-} \text{, and phosphite,} \left[\begin{array}{c} O \quad O \\ \| \quad \diagup \\ H - P \\ \diagdown \\ O \end{array} \right]^{=} \text{are also formed.}$$

The industrial method for preparing hypophosphite is based on the same reaction (see Chapter 12). In other words, it is not possible with this reaction to produce phosphine without the hypophosphite and vice-versa. The following equation of the hydrolysis of phosphorus with sodium hydroxide shows only the major possible reactions which occur.

$$3 \text{ NaOH} + 4 \text{ P} + 3 \text{ H}_2\text{O} \longrightarrow \quad 3 \text{ NaH}_2\text{PO}_2 \quad + \quad \text{PH}_3$$

sodium   phosphine
hypophosphite

Phosphine Flame Retardants for Cotton Fabrics. THPC. The research which led to the use of phosphine as an intermediate for the preparation of a flame-retarding composition is an interesting story. The original research was carried out by Wilson Reeves and John Guthrie at the Southern Regional Research Laboratory of the U.S. Department of Agriculture in New Orleans. John told me how he and Wilson arrived at their process. During the Korean conflict in the early 1950's, a research program was initiated at their laboratory to develop a process for flameproofing cotton for use in military clothing. Based on his previous knowledge, Dr. Guthrie felt that new flameproofing compounds should not wash off. Further, they should not be ionic—*i.e.*, they should not pick up sodium and calcium ions from laundering solutions. In the literature he found that trimethylene trisulfone, $(CH_2SO_3)_3$, reacts at room temperature with an alkaline water solution of formaldehyde to form a water-insoluble polymer. If this polymer were deposited on a fabric, it might impart permanent flameproofing.

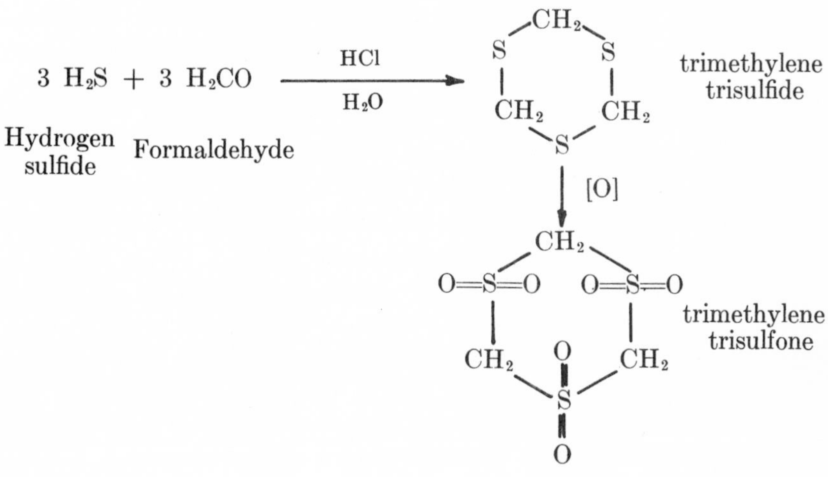

When he carried out his experiment, however, the treated fabric had only low flame resistance. It also lacked glow resistance—that is, the char which formed on burning continued to burn by afterglow. Guthrie knew that this afterglow did not happen to fabrics flameproofed by phosphorus compounds. His next thought was to synthesize a phosphorus analog of trimethylene trisulfone to see if it would form a phosphorus-containing polymer on reaction with formaldehyde. Since trimethylene trisulfone is made from trimethylene trisulfide, $(CH_2S)_3$, the first step, was to make a phosphorus analog of trimethylene trisulfide.

Trimethylene trisulfide is made by passing hydrogen sulfide (H_2S) gas into a water solution of formaldehyde (H_2CO) and hydrochloric acid (HCl). It is then oxidized to trimethylene trisulfone. This sequence of reactions is depicted in the equations at the bottom of page 109.

Since no phosphorus analog of trimethylene trisulfide was reported in the literature, Guthrie assumed that a method analogous to that for trimethylene trisulfide preparation would work. Phosphine, PH_3, (or hydrogen phosphide, the phosphorus analog of hydrogen sulfide) would be passed into a water solution of formaldehyde and hydrochloric acid. The reaction envisioned is:

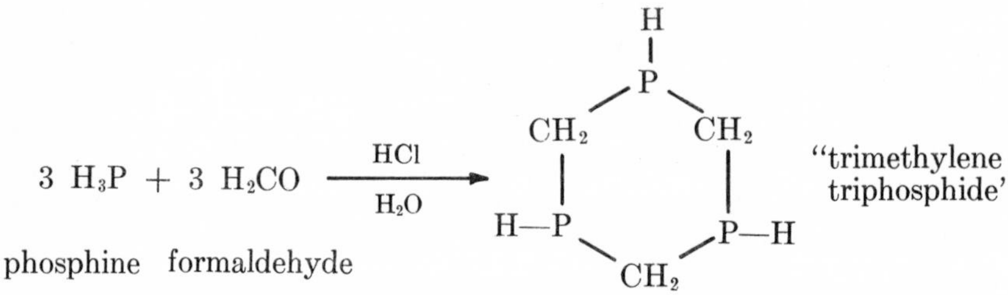

$$3 \ H_3P + 3 \ H_2CO \xrightarrow[\text{H}_2\text{O}]{\text{HCl}}$$

phosphine   formaldehyde

"trimethylene triphosphide"

However, when this reaction was carried out, "trimethylene triphosphide" was not found. The reaction mixture was dried, and a white crystalline solid was obtained. This solid was identified as tetrakis(hydroxymethyl)-phosphonium chloride, $[(HOCH_2)_4PCl]$, an unwieldy name soon shortened to THPC. A literature check showed that THPC had been made in 1921 by Alfred Hoffman, an independent American research chemist:

$$H_3P + 4 \ H_2CO + HCl \xrightarrow{H_2O} (HOCH_2)_4PCl$$

THPC

Since THPC was not the compound Guthrie and Reeves were looking for, less persistent and astute workers might have quit. However, Wilson Reeves, who did the actual experimental work, didn't quit. He tested the reaction of THPC with various classes of chemicals.

At the same time, in connection with another project, Guthrie had made some partially aminoethylated cotton in which some of the hydroxy (–OH) groups in the cellulose molecule were replaced by an aminoethyl group $(-OCH_2CH_2NH_2)$. Reeves found that THPC reacted with the amino group $(-NH_2)$ of the aminoethylated cotton and was thus chemically bonded to the cotton. The first step of this reaction is:

$$\text{CellO—CH}_2\text{CH}_2\overset{\overset{\displaystyle H}{|}}{\text{N}}\underset{\text{-------}}{[\text{H}} + \text{HO}]\text{CH}_2\text{P}^+(\text{CH}_2\text{OH})_3 \quad \text{Cl}^- \longrightarrow$$

aminoethylated
cellulose   THPC

$$\text{CellO—CH}_2\text{CH}_2\overset{\overset{\displaystyle H}{|}}{\text{N}}\text{—CH}_2\underset{\text{Cl}^-}{\text{P}^+}(\text{CH}_2\text{OH})_3 + \text{H}_2\text{O}$$

Here, a cellulose molecule was connected chemically, though indirectly, to a phosphorus compound. Since the phosphorus compound was not an ammonium salt and would not exchange sodium and calcium ions upon laundering, it wouldn't wash off. A fabric containing such treated cellulose molecules should also have some flame resistance. This was indeed the case. The project was a success, theoretically. Practically, it was not. The intermediate step, the joining of the aminoethyl group to the cellulose molecules in the cotton fabric, was simply too expensive. Chemically, THPC can react directly with cellulose to yield a phosphorus derivative of cellulose. However, the conditions necessary for this reaction are so drastic that the fabric is tenderized. Although clothing made from such fabric is flame resistant, it tends to fall apart.

In recent years, another treatment for cotton fabric was developed that imparts a wash-and-wear quality (permanent press) to the finished garment. In the early days of this treatment methylol urea,

$$(\text{HOCH}_2\overset{\overset{\displaystyle H}{|}}{\text{N}}\text{—}\overset{\overset{\displaystyle O}{\|}}{\text{C}}\text{—}\overset{\overset{\displaystyle H}{|}}{\text{N}}\text{CH}_2\text{OH}),$$ and/or methylol melamine were used:

$$\left[\begin{array}{c} \text{HOCH}_2\overset{\overset{\displaystyle H}{|}}{\text{N}} - \overset{\displaystyle N}{\underset{\displaystyle | \\ N \diagdown \quad \diagup N \\ | }{\diagup \quad \diagdown}} - \overset{\overset{\displaystyle H}{|}}{\text{N}}\text{CH}_2\text{OH} \\ \text{H—NCH}_2\text{OH} \end{array} \right]$$

methylol melamine

These compounds are made from urea or melamine and formaldehyde. The cloth is dipped in a water solution of these chemicals, dried, and cured at about 330°–340°F. During treatment the methylol groups (–CH$_2$OH) react with the hydroxyl groups (–OH) in the cellulose molecule. Since each molecule of methylol urea has two methylol groups ($-CH_2OH$), it can link two cellulose molecules together. Accordingly, each methylol melamine contains three methylol groups, so it can link three cellulose molecules together. When the cellulose molecules are thus crosslinked, the fibers become more rigid and harder to crease, or, when creased, easier to smooth again. This property is wrinkle resistance.

The above discussion shows that the methylol group ($-CH_2OH$) in methylol urea or in methylol melamine reacts with the hydroxyl group ($-OH$) of cellulose and is chemically bonded to it by the following reaction:

$$CellOH \ + \ HOCH_2\overset{\overset{\text{H}}{|}}{N}-\overset{\overset{\text{O}}{||}}{C}-\overset{\overset{\text{H}}{|}}{N}-CH_2OH \ \longrightarrow$$

$$CellO-CH_2\overset{\overset{\text{H}}{|}}{N}-\overset{\overset{\text{O}}{||}}{C}-\overset{\overset{\text{H}}{|}}{N}CH_2OH \ + \ H_2O$$

Reeves found that THPC, besides being able to react with the amino group of aminoethyl cellulose, can also react with the $-NH_2$ group in other compounds: for example, the H_2N- groups in urea,

($H_2N-\overset{\overset{\text{O}}{||}}{C}-NH_2$). He also found that THPC reacts with methylol groups ($HOCH_2-$) in methylol urea, ($HOCH_2\overset{\overset{\text{H}}{|}}{N}-\overset{\overset{\text{O}}{||}}{C}-\overset{\overset{\text{H}}{|}}{N}CH_2OH$), and also the methylol groups in methylol melamine. The reaction of ($HOCH_2)_4PCl$ occurs through the three methylol groups, ($HOCH_2-$). The fourth $HOCH_2-$ group splits off with the Cl group as HCl and formaldehyde (H_2CO). Since there are three reactive methylol groups in THPC, two reactive H_2N- groups in urea, and three reactive methylol ($HOCH_2-$) groups in trimethylol melamine, the whole mixture should be able to react together to form a polymer. It does.

If the ratio of reactants used is controlled so that there are some methylol groups ($HOCH_2-$) left from the methylol melamine, these should then react with the hydroxyl ($-OH$) groups of the cellulose molecule in the cotton. The treated cotton should then withstand laundering. This hypothesis was experimentally confirmed. The first flame retardant formulation developed consisted of a water solution of THPC, urea, methylol melamine, and triethanolamine. Since triethanolamine is a base, it ties up the HCl liberated from the THPC. After the THPC flame retardant finish is chemically bonded to the cloth, it is further treated with a hydrogen peroxide solution to oxidize the phosphine ($\geqslant$P) in the finish to phosphine oxide ($\geqslant$P $=$ O). This process is idealized as:

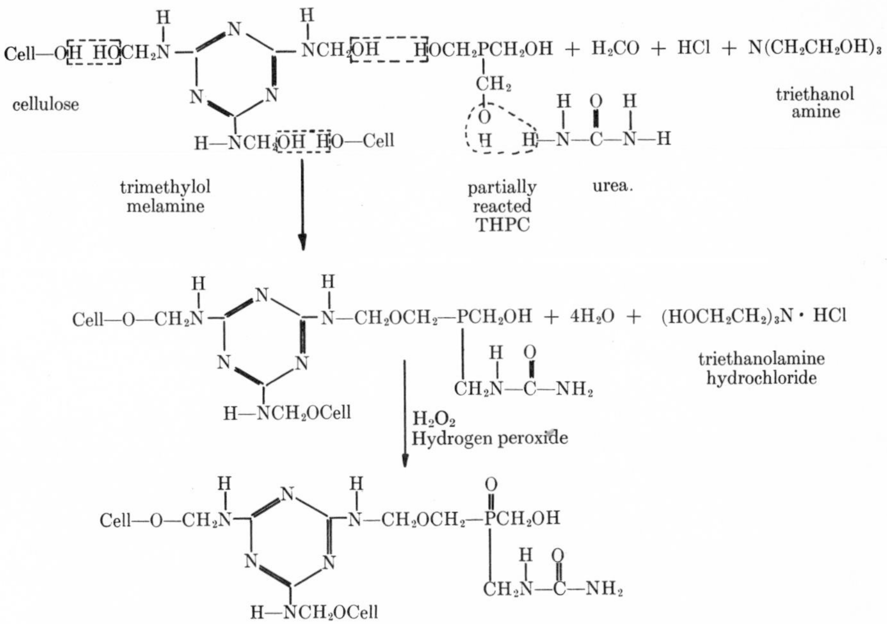

Continued modifications of the original THPC process showed that the treated cloth could be cured in the presence of gaseous ammonia (NH$_3$) at room temperature. This curing at low temperature preserves the original strength of the fabric. The excess ammonia also takes up the liberated hydrogen chloride and converts it to ammonium chloride. Curing with ammonia joins the methylol groups through the carbon-to-nitrogen-to-carbon linkage (—C—N—C—) in place of the carbon-to-oxygen-to-carbon linkage (—C—O—C—) shown previously. This development is an improvement over the original process (*see* top of page 114).

Recent U.S. and British law requires certain wearing apparel, especially children's nightwear, to be flame resistant. The modified THPC process was used extensively for this purpose. British regulations are particularly strict because many of their houses are still heated by fireplaces which burn "solid fuel" (a British term usually referring to coal). On cold mornings, children in long flannel nightgowns snuggle close to the fireplace, and their nightgowns often catch fire. Many children have been seriously injured or burned to death.

Even though the THPC flameproofing process is good, it has drawbacks; the treated cloth tends to be stiff, it crinkles like paper, and the treatment is expensive. According to Dr. J. W. Weaver, former research manager for Cone Mills Corp., the 1971 cost of material for children's sleepwear was about 27¢ per yard. The flame-retardant process added another 20¢ per yard. A garment made of flame-retardant material might retail for $3.98, for example, while the same garment, untreated, could be sold at about half that price.

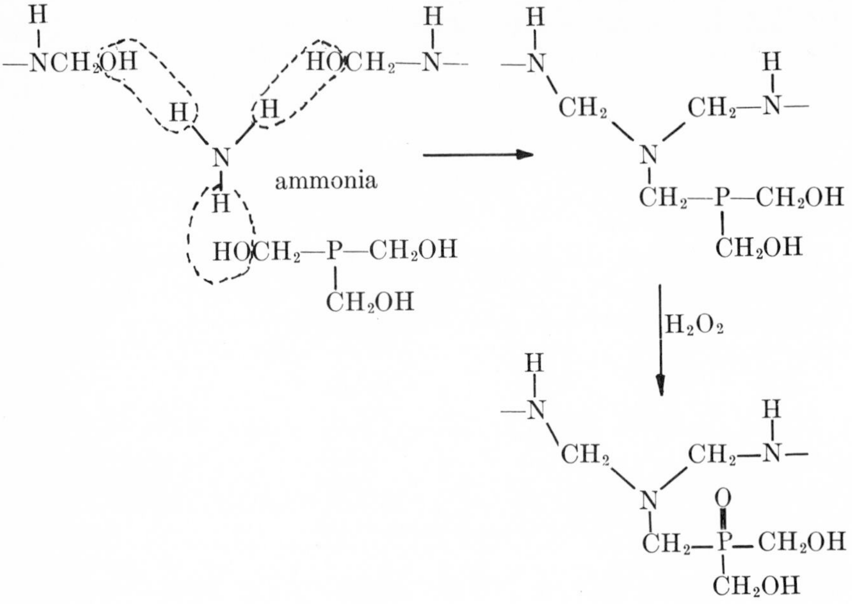

Some mothers believe that fire accidents happen only to other people's children. I have been told that since by law only flame-retarded ready made night garments are available for purchase in the stores, British housewives try to save money by buying untreated flannelette and making the garments themselves.

THPOH. Research in this area is continuing. In 1967, Wilson Reeves and his co-workers announced the THPOH process. [The formula for THPOH, or tetrakis(hydroxy) phosphonium hydroxide, is $(HOCH_2)_4POH.$] In this process, THPC is replaced by THPOH. THPOH is prepared from THPC by reaction with sodium hydroxide under carefully controlled conditions. This reaction is shown below.

$$(HOCH_2)_4PCl + NaOH \longrightarrow (HOCH_2)_4POH + NaCl$$
$$\quad\text{THPC} \qquad\qquad\qquad\qquad \text{THPOH}$$

Except for the replacement of THPC by THPOH, the basic chemistry of two processes is essentially the same. Ammonia is used to cure the polymer onto the fabric. The new process is claimed to give a flame-resistant cloth which retains most of its original strength and is about as soft as the untreated fabric. In 1974 this process had reached commercial production.

Organic Phosphonates

Pyrovatex CP. A more recent flame-retardant process which has reached commercial scale production is called Pyrovatex CP. According to the patent literature its phosphorus-containing component is:

$$(CH_3O)_2 \overset{\overset{\displaystyle O}{\|}}{P} CH_2CH_2 \overset{\overset{\displaystyle O}{\|}}{C} - \overset{\overset{\displaystyle H}{|}}{N} CH_2OH$$

[3-[(hydroxymethyl)amino]-3-oxopropyl]phosphonic
acid, dimethyl ester

The chemistry involved in attaching this molecule to cellulose is the same as that described for the THPC process—*i.e.*, through the methylol group, –CH₂OH. This group can react directly with the –OH groups of the cellulose, or it can bond indirectly to the cellulose *via* such compounds as methylol urea or methylol melamine. Cotton fabric properly treated

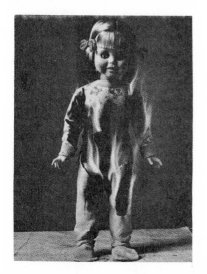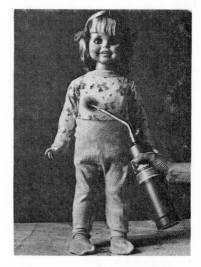

Courtesy CIBA-GEIGY Corp.

Untreated cotton flannel bursts into flame and burns vigorously. The same sleepwear treated with Pyrovatex chars but does not burn.

with Pyrovatex has a very good "hand" or softness. Its flame retardance withstands more than 50 machine launderings with phosphate detergents. However, laundering with non-phosphate detergents completely destroys its flame retardance. This could be caused by the high alkalinity of the non-phosphorus detergents, which tend to decompose the phosphorus compound. It may also be caused by the fact that non-phosphate detergents soften water by precipitating Ca^{2+} and Mg^{2+} cations as salts, whereas phosphate detergents remove Ca^{2+} and Mg^{2+} as water-soluble phosphate complexes. Residual precipitates of calcium and magnesium salts on the fabric, upon pyrolysis, would form non-flame-retardant phosphate salts rather than flame-retardant phosphoric acids.

Fyrol 76. Another flame-retardant process which reached commercial status in 1974 is called Fyrol 76 after the composition of the same name which contains a vinylphosphorus $(CH_2{=}CH{-}\overset{\overset{\displaystyle O}{\|}}{P})$ group. This compound bonds to cellulose also *via* a methylol group but in this case, it does so indirectly *via* the methylol group of methylol acrylamide. A free radical catalyst, potassium persulfate, $K_2S_2O_8$ is used to join the double bonds of the vinylphosphate with that of the acrylamide. The methylol group is attached to the cellulose when the treated fabric is heated at 300°–350°F:

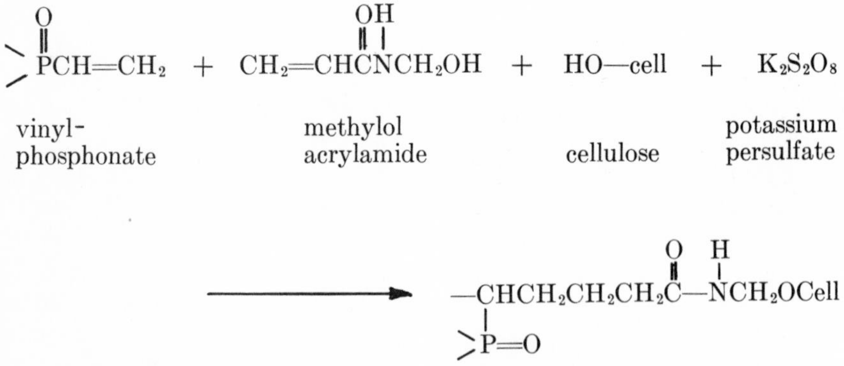

$$\overset{\overset{\displaystyle O}{\|}}{\underset{}{P}}CH{=}CH_2 \;+\; CH_2{=}CH\overset{\overset{\displaystyle OH}{\|\;|}}{C}NCH_2OH \;+\; HO{-}cell \;+\; K_2S_2O_8$$

vinyl-phosphonate   methylol acrylamide   cellulose   potassium persulfate

$$\longrightarrow \; {-}CHCH_2CH_2CH_2\overset{\overset{\displaystyle O}{\|}}{C}{-}\overset{\overset{\displaystyle H}{|}}{N}CH_2OCell$$
$$\underset{}{>}P{=}O$$

Flame retardance from Fyrol 76 with a 20–25% add-on by weight on lightweight cotton flannel withstands more than 50 launderings. The treated fabric exhibits soft hand and good strength. However, like the Pyrovatex and THPC treatments, the treated fabrics should not be laundered in non-phosphate detergents built with sodium carbonate, or in soap in hard water.

Fyrol 6. Phosphorus compounds containing halogens and nitrogen are also used to synthesize flame retardants. For example, phosphorus trichloride reacts with ethyl alcohol, C_2H_5OH, to give diethyl phosphonate:

$$PCl_3 + 3\ C_2H_5OH \longrightarrow (C_2H_5O)_2\overset{\displaystyle O}{\overset{\|}{P}}H + C_2H_5Cl + 2\ HCl$$

ethyl   diethyl   ethyl

alcohol   phosphonate chloride

Although diethyl phosphonate has been patented as a solvent in paint removers, its largest use is as an intermediate in preparing a reactive flame retardant for rigid polyurethane foam. The flame retardant is made by the reaction of diethyl phosphonate, $(C_2H_5O)_2\overset{O}{\overset{\|}{P}}H$, with formaldehyde, CH_2O, and diethanolamine, $HN(CH_2CH_2OH)_2$. The resulting compound is Fyrol 6.

$$(C_2H_5O)_2\overset{\displaystyle O}{\overset{\|}{P}}H + CH_2O + HN(CH_2CH_2OH)_2 \longrightarrow$$

diethyl   formaldehyde   diethanolamine

phosphonate

$$(C_2H_5O)_2\overset{\displaystyle O}{\overset{\|}{P}}CH_2N(CH_2CH_2OH)_2$$

diethyl *N,N*-bis(hydroxy-
ethyl)aminomethylphosphonate
Fyrol 6

Rigid polyurethane foams are used extensively for packaging and insulation, particularly in household refrigerators, refrigerated railway cars ("reefers"), refrigerated trucks, and insulating wall boards. As one of the newer insulating materials, urethane foam is so efficient that new refrigerators have thinner walls and thus more inside space. When Fyrol 6 is added to the polyurethane formulation, it reacts with the diisocyanate during the foam formation stage to become an integral part of the foam. The resulting foam, with approximately 1.0–1.5% phosphorus, is flame resistant.

Although Fyrol 6 is an effective flame retardant for rigid urethane foam, it is not suitable for flexible urethane foam because it imparts undesirable physical properties to the latter. However, flexible urethane foams are used extensively as cushions in automobiles, furniture, and mattresses. Without proper flame retardants, these foams burn quite readily. As more cotton batting cushioning materials are replaced by flexible urethane foams, this problem becomes more serious. Even cotton batting cushions, when not properly flame retarded, are the source of many serious fires—e.g., burning of mattresses by people who fall asleep while smoking in bed.

Organic Phosphates

Phosgard 2XC20. One phosphorus-containing flame retardant used commercially in flexible urethane foams—as well as in rigid urethane foams and other polymers—goes by the tradename of Phosgard 2XC20. It too is made from phosphorus trichloride:

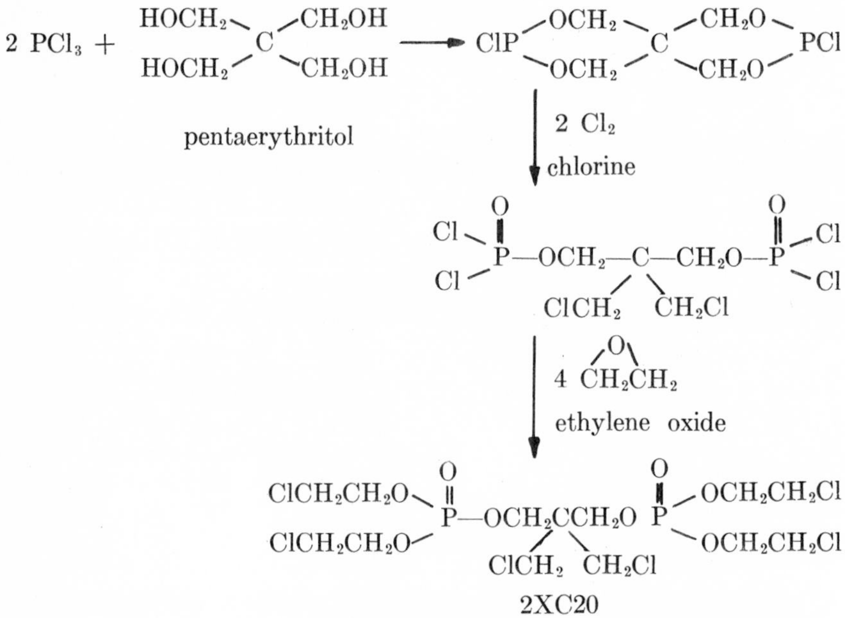

pentaerythritol

2,2-bis(chloromethyl)-1,3-propanediol
bis [bis(chloroethyl) phosphate

Even though the resulting chemical structure seems fairly complicated, the starting materials are quite common. The tetrahydric alcohol used—pentaerythritol—is an ingredient used to prepare alkyd resins for enamels applied to stoves, refrigerators, and washing machines.

Phosgard 2XC20, unlike Fyrol 6, is not a reactive flame retardant. It is incorporated into the system as an additive. Like all additive flame retardants in general, when subjected to elevated temperatures such as those encountered in automobile cushions while under a hot sun, a certain amount of it volatilizes. Depending on the additive, some volatilize more than others. These volatilized additives, when condensed on the colder windshield and windows, cause fogging which obscures vision.

Polyurethane foam formulated with Stauffer's Fyrol 6 flame retardant is sprayed on the roof of the Superdome in New Orleans

In order to be acceptable for automotive foams, the volatility has to be low enough to meet the requirements of the automobile manufacturers.

Fyrol FR-2. Tris(dichloroisopropyl) phosphate, or Fyrol FR-2 is another important flame retardant for flexible urethane foam. It is prepared by the reaction of phosphorus oxychloride with epichlorohydrin.

$$POCl_3 + 3\ ClCH_2CH\overset{O}{-}CH_2 \longrightarrow O{=}P\left(OCH\overset{CH_2Cl}{\underset{CH_2Cl}{\diagup}}\right)_3$$

phosphorus   epichloro-
oxychloride   hydrin

tris(dichloroisopropyl)
phosphate(Fyrol FR-2)

The product is a phosphate ester and is formed by the opening of the epoxide ring,

$$-\overset{\displaystyle O}{\overset{\displaystyle /\ \ \backslash}{\underset{}{CH-CH-}}},$$

of the epichlorohydrin by the PCl group of phosphorus oxychloride.

Fyrol CEF, Union Carbide "3CF." Another compound which is used as a flame retardant for flexible urethane as well as rigid urethanes and other polymers—usually in combination with other flame retardants—is tris(β-chloroethyl) phosphate. The chemistry for its preparation is also based on the opening of the epoxide ring with phosphorus oxychloride:

$$POCl_3 + 3 \ \overset{\displaystyle O}{\overset{\displaystyle /\ \ \backslash}{CH_2-CH_2}} \quad\longrightarrow\quad OP(OCH_2CH_2Cl)_3$$

tris(β-chloroethyl) phosphate

Firemaster T23P, Fyrol HB-32. A particularly effective flame retardant which is also derived from phosphorus oxychloride is tris(dibromopropyl) phosphate (TDBPP). It is known by such tradenames as T23P and HB-32.

$$POCl_3 + 3 \ CH_2BrCHBrCH_2OH \rightarrow OP(OCH_2CHBrCH_2Br)_3 + 3 \ HCl$$

The effectiveness of this compound as a flame retardant is enhanced by the presence of bromine, a known efficient flame retardant. As shown in the reaction, this compound is formed by the direct esterification of the bromine-containing alcohol with phosphorus oxychloride. TDBPP is used in flexible and rigid polyurethane foams. It is also incorporated as an additive in solutions of cellulose polymers or acrylic polymers before they are extruded into fibers. These fibers are then used to make flame-retardant clothing, draperies, curtains, carpets, and even doll's hair. In a very recent application TDBPP is used to impart some flame retardance to polyester fabric. In one technique, a water emulsion of TDBPP is added on the polyester fabric. After drying, the treated fabric is heated to about 400°F for one minute. Under these conditions, some TDBPP is absorbed into the fabric and is stable enough to resist a certain amount of laundering.

In the United States, most flame-resistant garments are made for the military and for factory workers. Workers in steel mills who are constantly exposed to flying sparks and molten metals prefer the flameproof garments—stiffness, crinkle, higher cost, and all. Flame-retardant garments for sleepwear are also appearing on the market. Our concern with flame retardance is so great because studies show that every year about 3000 to 5000 deaths and 150,000 to 250,000 injuries are caused by burns in connection with flammable fabrics—in garments worn by millions of people who do not realize their potential hazard. It was the infamous torch sweaters which brought about the passage of the Flammable Fabric Act of 1953; this Act was amended in 1967. The Department of Commerce has banned the sales of flammable sleepwear for children after July 1973. It has also issued regulations covering carpets and mattresses. The Department of Transportation has issued flammable rules for automobile interiors and has upgraded regulations for passenger airplane interiors. The Department of Health, Education, and Welfare (HEW) has set flammability standards on interior furnishings of those hospital and nursing homes built under the provisions of the Hill Barton Act.

Flame-retarding processes for many systems are available, but not without shortcomings. For example, for clothing materials we still need a process which is easy to apply, economical, which will not adversely affect the strength, feel, and other properties of the fabric, and which is permanent to laundering in phosphate and non-phosphate detergents. In other words, the development of a really good flame-retardant process for fabrics is still a challenge for scientists. Even more challenging is the development of a flame-retardant process for the widely used polyester–cotton blends. Pure polyester fabrics which are not flame retarded burn by melting, and the flame drips off with the melt. The burns from this flame and hot melt are painful, but such injuries are not as severe as those which result when the whole garment catches on fire. Extensive studies by HEW showed that there is a remarkable correlation between

Protective workclothes and children's sleepwear are key markets for fire-retardant chemicals. Increased concern with fire-related deaths, particularly among children, has led to increasing legislation requiring that garments pass certain fire-retardancy standards. The most stringent tests currently apply only to children's sleepwear.

the extent of the injury to the victim's body and the distribution of the burn on the clothing.

For pure cotton fabrics, at least, several flame retarding treatments do exist (with their merits and drawbacks). For polyester–cotton blends, however, even if the cotton portion is made flame-retardant, the blend still burns; in fact, it burns more dangerously than pure polyester because the char from the burned cotton supports the burning polyester portion, preventing it from dripping away. Some flame retardants for polyester–cotton fabrics are under development in various laboratories. However, as of 1974 they still have enough drawbacks to preclude their introduction on a commercial scale.

11

Water Softening with Phosphates

Ground water from deep wells which is used for drinking, laundry, bathing, or for generating steam in a power plant is called hard water because it contains an appreciable amount of calcium and magnesium ions, Ca^{2+} and Mg^{2+} (also iron, Fe^{2+}). These ions are usually present as the soluble calcium and magnesium hydrogen carbonates [$Ca(HCO_3)_2$ and $Mg(HCO_3)_2$] and magnesium sulfate ($MgSO_4$). They react with soap to form insoluble salts which are deposited as a bathtub or sink ring. When some hard waters are heated, a precipitate will appear as scale inside the heating container. Why does this scale form? The answer depends on the salts in the water. For example, when a metal hydrogen carbonate is heated, it decomposes to liberate CO_2 and forms the metal carbonate:

$$M(HCO_3)_2 \xrightarrow{\text{heat}} MCO_3 + CO_2 + H_2O$$

Thus, in hot water heaters or in steam-generating plants, the soluble calcium and magnesium bicarbonates decompose into the insoluble calcium and magnesium carbonates which appear as a hard scale. Some of the water evaporates, and calcium sulfate may also come out of solution. The scales thus formed decrease the efficiency of the heater since they are poor heat conductors.

Softening Water with Trisodium Phosphate

Trisodium phosphate reacts with calcium and magnesium ions in water to form insoluble tricalcium and trimagnesium phosphates. When these ions are removed, the water is softened.

In practice, trisodium phosphate is used only in conjunction with other, cheaper water-softening processes, such as the soda-lime process, to precipitate residual calcium and magnesium ions as the very insoluble hydroxyapatite and trimagnesium phosphate. The precipitate is dispersed and does not form a hard scale. Therefore, it is easily purged. The reaction for removing calcium and magnesium from boiler water is:

$$5\,Ca(HCO_3)_2 + 4\,Na_3PO_4 \xrightarrow{H_2O} Ca_5(OH)\,(PO_4)_3 + 10\,NaHCO_3 + Na_2HPO_4$$

hydroxyapatite

$$3\,MgSO_4 + 2\,Na_3PO_4 \xrightarrow{H_2O} Mg_3(PO_4)_2 + 3\,Na_2SO_4$$

trimagnesium
phosphate

Metal hydrogen carbonate also is converted to the metal carbonate when neutralized with a base. In the soda-lime process hard water is first treated with sodium carbonate and then with lime to convert the soluble calcium and magnesium bicarbonates into the relatively insoluble calcium and magnesium carbonates.

The Many Classes of Sodium Metaphosphates (Sodium Polyphosphates)

Sodium metaphosphates represent one of the most important and interesting classes of phosphates. All three sodium metaphosphates described here are made from the same raw material. Since their chemical compositions are nearly identical, one would expect their properties to be similar, but this is not the case. For example, one compound is extremely soluble in water and is a very good water softener. Another is insoluble in water and is used as a polishing agent in toothpastes (Chapter 5). The third has an interesting ring structure and is used to modify starch used in making clear, thick soups as described in Chapter 4. How can the same raw material be converted to products with such contrasting properties? Why are they useful? What research was done to put these compounds to work?

The term metaphosphate refers to compounds with one or more

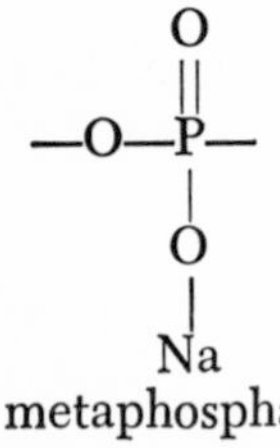

$$\begin{array}{c} O \\ \parallel \\ -O-P- \\ | \\ O \\ | \\ Na \end{array}$$

units. Since the polyphosphates are chains built up from metaphosphate units, they are sometimes called polymetaphosphates or simply polyphosphates. The term sodium hexametaphosphate is also used. It implies that the compound is a polyphosphate chain containing six (hexa-) metaphosphate ($NaOPO_2$) units. However, research has shown this notion to be incorrect. Commercial sodium hexametaphosphate actually contains an average of 13–18 metaphosphate units.

At present, there are three sodium metaphosphates that are important in industry: glassy sodium polymetaphosphate, cyclic sodium trimetaphosphate, and the insoluble sodium metaphosphate (Chapter 5). All are prepared from the same material—monosodium phosphate—and it is the reaction conditions that determine which product is formed. The preparation of cyclic sodium trimetaphosphate is decribed in detail in Chapter 4.

Glassy Sodium Polymetaphosphate. As described in Chapter 4 sodium acid pyrophosphate is prepared by heating monosodium phosphate at about 225°–250°C, thus condensing two moles of monosodium phosphate together and eliminating one mole of water:

$$
\underset{\substack{\text{monosodium}\\\text{phosphate}}}{
\begin{array}{c}
\quad\;\; O \qquad\qquad O \\
\quad\;\; \| \qquad\qquad \| \\
NaOPOH \; + \; HOPONa \\
\quad\;\; | \qquad\qquad\;\; | \\
\quad\;\; O \qquad\qquad O \\
\quad\;\; | \qquad\qquad\;\; | \\
\quad\;\; H \qquad\qquad\; H
\end{array}}
\xrightarrow{225°-250°\ C}
\underset{\substack{\text{sodium acid}\\\text{pyrophosphate}}}{
\begin{array}{c}
O \qquad\;\; O \\
\| \qquad\;\; \| \\
NaO-P-O-P-ONa \\
\;\; | \qquad\qquad | \\
\;\; O \qquad\qquad O \\
\;\; | \qquad\qquad | \\
\;\; H \qquad\qquad H
\end{array}}
+ \; \underset{\text{water}}{H_2O}
$$

If monosodium phosphate is heated at 800°–900°C rather than at 225°–250°C, the reaction mixture becomes a hot molten liquid, and almost all water is eliminated. If the molten liquid is cooled quickly (*e.g.*, by pouring over a cold steel sheet), the material solidifies into a thin layer of colorless glass. This glassy product is sodium polymetaphosphate, or Graham's salt (after Thomas Graham, who first described it in 1833). The size of each molecule depends on the ratio of sodium to phosphorus used. The reaction is:

$$
\underset{\text{monosodium phosphate}}{
\begin{array}{c}
\;\; O \qquad\qquad\quad O \qquad\qquad\quad O \\
\;\; | \qquad\qquad\quad \| \qquad\qquad\quad \| \\
HOPOH \; + \; n\,(HOPOH) \; + \; HOPOH \\
\;\; | \qquad\qquad\qquad | \qquad\qquad\qquad | \\
\;\; O \qquad\qquad\quad O \qquad\qquad\quad O \\
\;\; | \qquad\qquad\quad\; | \qquad\qquad\quad\; | \\
\;\; Na \qquad\qquad\;\; Na \qquad\qquad\;\; Na
\end{array}}
$$

$$
\xrightarrow{\hspace{2cm}}
\underset{\substack{\text{glassy sodium}\\\text{polymetaphosphate}}}{
\begin{array}{c}
\;\; O \quad\;\; O \quad\;\; O \\
\;\; \| \quad\;\; \| \quad\;\; \| \\
HOP-O-P-O-POH \\
\;\; | \qquad | \qquad | \\
\;\; O \qquad O \qquad O \\
\;\; | \qquad\; | \qquad\;\; | \\
\;\; Na \quad\; Na \; _n\; Na
\end{array}}
+ \; (n+1)\,H_2O
$$

The n in the equation represents the many metaphosphate units in the chain other than the end units. Obviously, the larger the value of n, the longer the chain.

Cyclic Sodium Trimetaphosphate. If monosodium phosphate is heated to about 530°–600°C, it will not go into a molten form. Three moles of monosodium phosphate will join head-to-tail and lose three moles of water to form a ring. The compound formed is called cyclic trimetaphosphate and is a white crystalline product. Various other methods for its preparation have been described in Chapter 4.

IMP. Insoluble sodium metaphosphate (IMP) is sometimes known as Maddrell's salt, after the German chemist, R. Maddrell. There are several forms of IMP, but the industrially important one is generally made by heating monosodium phosphate at about 475°–500°C. The crystalline product which is almost insoluble in water, forms readily in small quantities. In fact, in preparing cyclic trimetaphosphate or sodium acid pyrophosphate, using monosodium phosphate as the raw material, IMP forms as a by-product if reaction conditions are not controlled. The presence of even 0.1% IMP as by-product renders a water solution of the desired water-soluble compound cloudy and turbid. Sometimes IMP even shows up as the undesirable by-product in sodium tripolyphosphate.

Why is IMP insoluble in water when other sodium metaphosphates are very soluble? X-ray diffraction studies show that an IMP crystal is two long metaphosphate chains which spiral in opposite directions around the skew axes of the crystal. These chains are held together so tightly that it is difficult for a water molecule to slip in. However, if some of the sodium ions in IMP are exchanged with larger potassium ions, the compactness is destroyed and IMP gradually dissolves. An interesting experiment is to add a mixture of IMP and potassium metaphosphate to distilled water. Neither compound alone is water soluble, but the mixture gradually dissolves because some sodium and potassium ions have exchanged from one compound to the other.

IMP must be manufactured under static conditions or some soluble metaphosphate forms. The reason may be that since IMP's insolubility is the result of a compact crystal structure, this structure must be allowed to grow undisturbed or the structure will be broken up.

Softening Water with Glassy Sodium Metaphosphate

The most popular and well-known tradename for glassy sodium polymetaphosphate is Calgon, now a household word. Although this compound has been known since 1833 when Graham first described it, it was not until 1929 that Ralph E. Hall discovered its unique property— the ability to sequester many metallic ions as soluble complexes. Hall found this property useful for softening water, and his discovery led to many other applications which formed the basis of a large industry. The name Calgon was coined by Dr. Hall when he observed that glassy sodium polymetaphosphate could sequester calcium ions in water and make it seem as if they were not there—*i.e.*, gone, or calcium gone, Calgon.

The theoretical value of P_2O_5 in a sodium metaphosphate unit ($NaOPO_2$) is 69.9%. If one uses the theoretical value for both the P_2O_5 content and the sodium oxide content—that is, 100% pure monosodium phospate as the raw material—the glassy polymetaphosphate obtained would be a long chain of infinite length. However, this is not practical. For industrial use, Calgon generally has a P_2O_5 content of around 67–67.8%, adequate for most applications. This amount is 1.8–2.6% below the theoretical value for sodium metaphosphate. This also means that the sodium oxide value is above the theoretical. In other words, the starting material is not pure monosodium phosphate, but monosodium phosphate with a little disodium phosphate.

Disodium phosphate $[(NaO)_2PO(OH)]$ acts as a chain terminator —*i.e.*, it becomes part of the end of the chain. Thus, when a polymetaphosphate chain reacts with disodium phosphate, the chain is terminated:

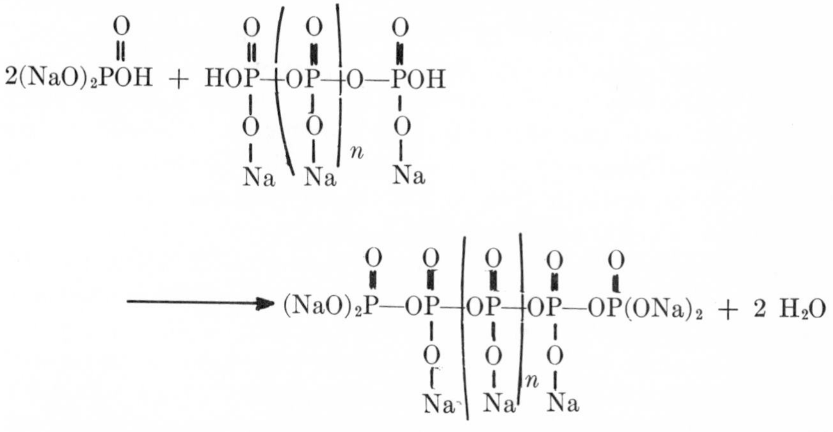

$$2(NaO)_2POH + HOP\!\left(\!OP\!\right)_n\!O\!-\!POH$$

$$\longrightarrow (NaO)_2P\!-\!OP\!\left(\!OP\!\right)_n\!OP\!-\!OP(ONa)_2 + 2\ H_2O$$

Of course, the more chain terminators there are, the shorter the chains will be. For commercial Calgon with 67–67.8% P_2O_5, the chain length is between 13 and 18 metaphosphate units. If there were more chain terminators (*e.g.*, if the total P_2O_5 used were lower), shorter chains would be possible. For instance, if the P_2O_5 content were down to 52.8%, the chain length would be only the two end units, or tetrasodium pyrophosphate. The chain length of commercial glassy sodium polymetaphosphate is adjusted to obtain good water solubility and less hygroscopicity as a solid.

Sodium polymetaphosphate was first used to soften water. It forms a soluble complex with the calcium and magnesium ions of hard water, and renders them unable to react with soap to form an insoluble scum. Even if an insoluble soap should form, glassy sodium polymetaphosphate can redissolve it. Some soap is always wasted by reaction with magnesium and calcium ions to form useless insoluble soaps, but if glassy sodium polymetaphosphate is added, no soap is wasted. Of course, if the water has already been softened, sodium polymetaphosphate is not needed.

Another important use for this compound is in the so-called "threshold" treatment. As described earlier, when hard water is heated in a boiler, the soluble calcium and magnesium salts originally present begin to precipitate out as calcium sulfate and calcium and magnesium car-bonates. These salts deposit as crystals (scales) on the inside surfaces of boilers. If these crystals are to form, nuclei must be present upon which more molecules can build in a regular and orderly fashion. Through this process, the nuclei grow into the large crystals known as scale. If anything is added to prevent this orderly growth, the crystal will not grow.

A small amount of glassy sodium polymetaphosphate interferes with crystal growth, thus preventing scale formation. It is not necessary to add so much sodium polymetaphosphate that all calcium ions and magnesium ions sequester into soluble complexes, just enough to prevent crystal growth. This is called the "threshold" treatment. If glassy sodium polymetaphosphate were added to sequester all the calcium, a glassy sodium polymetaphosphate-to-calcium ratio of 2.5 to 1 would be required. In many threshold treatments, 1 ppm of glassy sodium polymetaphosphate prevents the precipitation of calcium carbonate from a solution containing 200 ppm of calcium bicarbonate, $Ca(HCO_3)_2$.

Threshold treatment prevents the precipitation of scale in boilers. This allows the boiler to maintain high heat-transfer efficiency, but the treatment is not without drawbacks. A clean surface is also more prone to chemical attack and corrosion. Fortunately, the threshold treatment which prevents scale formation also protects against corrosion. A sub-microscopic protective film is formed by the reaction of the metal and metallic oxides on the inside surface of the boiler with the glassy sodium polymetaphosphate. This film insulates the clean metal surface from attack by oxygen and other corrosive elements in water. The film is so thin that it is visible only as a multicolored iridescence on a polished metal surface. In cold water, 2 ppm of Calgon will control corrosion. In hot water, 40–60 ppm are needed.

The threshold treatment also prevents iron or manganese compounds from precipitating when water is exposed to air or chlorine. Precipitation of ferric oxides gives red water. Precipitation of manganese compounds gives black water. In general, for each ppm of iron or manganese only 2–4 ppm of glassy sodium polymetaphosphate is needed. Threshold treatment is used in almost all industrial water supplies to prevent lime scales. It is useful in boilers, pipes, screens, washers, heat exchange equipment, and condensers.

Treatment of Hard Water with Tetrasodium Pyrophosphate

In addition to calcium and magnesium ions, tetrasodium pyrophosphate is also able to sequester other heavy metal ions such as iron and vanadium. In the textile industry many cottons are bleached by peroxy compounds such as hydrogen peroxide. However, peroxides are easily decomposed in the presence of even traces of heavy metal ions. It is therefore important to use TSPP and sodium silicate to sequester these metal ions, thus preventing them from catalytically decomposing the peroxides.

12

Electroless Plating

Many chrome-plated parts on new automobiles are not metal at all but chrome-plated plastic. Sodium hypophosphite is the compound usually responsible for this successful chrome plating. It is also used to plate corrosion-resistant nickel coatings on industrial machinery. Since all plating using sodium hypophosphite is done without an electric current, it is called electroless plating.

Sodium hypophosphite, as indicated by the prefix hypo-, has a valence of $+1$ (P^{1+}) which is lower than the valence of $+3$ (P^{3+}) of phosphite. It is made by the reaction of yellow phosphorus with a boiling water solution of a base such as lime [calcium hydroxide, $Ca(OH)_2$] or sodium hydroxide ($NaOH$). When lime is used, the products, in addition to water-soluble calcium hypophosphite, are insoluble calcium phosphite (removed by filtration) and the gaseous by-products, phosphine (PH_3) and hydrogen. The mechanism for this reaction is quite complicated and not perfectly understood. The following reaction shows only the overall process:

$$2\ P_4 + 4\ Ca(OH)_2 + 6\ H_2O \longrightarrow$$

phos-   calcium   water
phorus   hydroxide

$$2\ Ca(H_2PO_2)_2 + 2\ CaHPO_3 + 2\ H_2\uparrow + 2\ PH_3\uparrow$$

calcium   calcium   hydrogen   phos-
hypophos-   phosphite   gas   phine
phite   precipitate   gas

The water solution of calcium hypophosphite is converted to sodium hypophosphite by addition of sodium sulfate (Na_2SO_4). The by-product from this reaction—insoluble calcium sulfate—is removed by filtration. The solution is then concentrated to recover sodium hypophosphite crystals as the monohydrate ($NaO_2PH_2 \cdot H_2O$):

$$Ca(H_2PO_2)_2 + Na_2SO_4 \xrightarrow{\ H_2O\ } 2\ NaH_2PO_2 \cdot H_2O + CaSO_4$$

calcium   sodium   sodium   calcium sulfate
hypophosphite sulfate   hypophosphite   precipitate

Sodium hypophosphite electroless plating was first developed by General American Transportation Corp. under the trade name Kanigen for plating odd-shaped metal parts or the interiors of railroad tank cars with corrosion-resistant nickel coatings. The nickel coating produced also contains 8–10% phosphorus, apparently as phosphide. To apply this coating, a bath consisting principally of sodium hypophosphite and nickel sulfate ($NiSO_4 \cdot 6\ H_2O$) is used; the bath also contains a complexing agent to prevent the precipitation of nickel phosphite, accelerators to enhance the rate of nickel deposition, and stabilizers to prevent decomposition of the bath.

Many proprietary formulations on the market contain different auxiliary agents. The exact operating conditions for each bath depend on the nature of these added reagents. However, the main reaction for nickel plating involves the reduction of the nickel ion, Ni^{2+}, from nickel sulfate to nickel metal or Ni^0:

$$H_2PO_2^- + H_2O \longrightarrow H_2PO_3^- + 2\ H^+ + 2\ e^-$$

$$Ni^{2+} + 2\ e^- \longrightarrow Ni^{\ 0}$$

$$H_2PO_2^- + H_2O + Ni^{2+} \xrightarrow[\text{catalyst}]{Ni} H_2PO_3^- + 2\ H^+ + Ni^0$$

hypophos- phite ion	nickel ion	phosphite ion	nickel metal

Theoretically, electroless nickel plating with hypophosphite involves the same reaction as does plating with an electric current. That is, it requires the action of two electrons (e^-) to reduce the nickel ion of $+2$ valence (Ni^{2+}) to nickel metal (Ni^0) of zero valence. In electric plating, the two electrons are supplied by electric current. In electroless plating using sodium hypophosphite the two electrons come from the oxidation of the hypophosphite ($H_2PO_2^-$) to the phosphite ($H_2PO_3^-$). In other words, in oxidizing the hypophosphite ion to the phosphite ion, the phosphorus atom is changed from a valence of P^{1+} to P^{3+} with a loss of two electrons. This reaction is catalyzed by freshly deposited nickel and by such metals as palladium. Substrates to be plated such as iron or aluminum are placed in the plating solution; since these metals are more electropositive than nickel, some metallic nickel deposits, and simultaneously some iron or aluminum dissolves. This thin layer of deposited nickel then catalyzes the electroless plating of nickel on the iron or aluminum surface.

In many respects, electroless plating is more versatile than electroplating. Objects of complicated design, with many narrow recesses, crevices, and small holes cannot be completely coated in electroplating. The electric current jumps across narrow gaps, leaving the crevices and holes unplated and unprotected. In electroless plating, coverage occurs

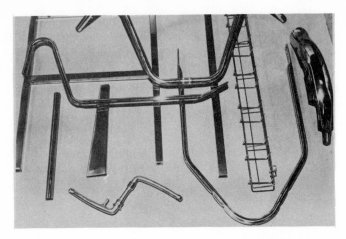

Courtesy Stauffer Chemical Co.

*These irregularly shaped parts were plated using phosphates
in the cleaner bath*

wherever the solution can reach. This results in complete, protective
coverage.

Small objects such as valve and pump parts made of steel become
quite corrosion resistant when nickel plated. The interiors of large tank
cars are very difficult to coat by electroplating. Electroless nickel plating,
on the other hand, works quite well; the entire tank car is used as the
plating bath. The ultimate goal of electroless plating is the formation of a
corrosion resistant coating on metals such as iron and aluminum. In the
chemical industry corrosion is the number one problem.

Chrome Plated Parts

Most automobile trim which is not made of bright-dip aluminum,
is fabricated of chrome-plated zinc die-cast pieces. The chrome is electro-
plated over a metal surface which has previously been plated (also
electrically) with copper and nickel. In fact, on some older automobiles—
including very expensive models—some of the shiny chrome plates are
marred by blisters. These are the result of imperfect plating which leads
to corrosion of the zinc die-cast metal base.

Plastic automobile parts resist the ravages of weather and salts used
on roads in winter. A chrome plate applied to a smooth plastic surface
imparts the desired decorative effect and yet will not blister from corrosion. In addition, plastic parts are easy to mold, and for this purpose
are cheaper than zinc die-cast metals; one reason is that, pound for pound,
plastic is much lighter than metal. As a result, many chrome-plated
plastic parts have replaced the commonly used zinc die-cast metal parts.

Plating of Plastics

One night while waiting to be served in a restaurant, I was idly playing with the salt shaker. The cap looked like any chrome-plated metal top, except that it was quite lightweight. Then I realized that it was made of chrome-plated plastic. Although I knew that many of the shiny parts in new automobiles (knobs, handles, trim, and housings for the lights) are chrome-plated plastic, I didn't realize that this technology had also invaded the household and restaurant markets.

Objects for electroplating must conduct electricity. Since plastics are non-conductors, they can be coated only by electroless plating of a nickel underlayer using a sodium hypophosphite–nickel sulfate bath. Electroless nickel plating of plastics is the result of much research. For many years only a specially developed plastic, known as ABS resin (acrylonitrile–butadiene–styrene), was suited to electroless plating. Now other plastics such as polypropylene have been reported to be platable. For successful plating, the coating must adhere tightly to the surface of an object. Since plastics are generally smooth, they present quite a challenge. Thus, a special type of plastic was developed whose surface could

Courtesy Borg-Warner Corp.

New technique for plating molded ABS plastic parts has been put into practical operation by large-scale commercial electroplaters. Picture taken in a Michigan plater's plant shows the electroless plating step, with ABS parts immersed in a special plating bath. This phase follows surface activation of the plastic. Next step will be a water rinse.

be treated to anchor the metal coating without seriously affecting the shine of the final chrome plate.

Here is how ABS is plated. The plastic piece is first etched with a sulfuric acid–chromic acid solution. The butadiene component in the plastic is attacked by this solution, and small pits are etched randomly on the surface. These pits are about one-half micron deep (1 micron = 0.001 millimeter) and have porous bases and sides. (To the naked eye, the plastic surface is still quite smooth.) The object is then placed in a bath containing stannous chloride ($SnCl_2$) and palladium chloride ($PdCl_2$). When these chemicals react, palladium chloride is reduced to palladium metal (Pd^0), and stannous chloride is oxidized to stannic chloride ($SnCl_4$). Palladium metal is deposited on the porous surface of the etched pits. This oxidation–reduction reaction is shown as follows:

$$SnCl_2 \quad + \quad PdCl_2 \quad \longrightarrow \quad Pd^0 \quad + \quad SnCl_4$$

stannous chloride	palladium chloride	palladium metal	stannic chloride

Since the palladium metal deposits only on the inside of the random pits, the plastic surface is still not conductive enough for electroplating. The plastic piece is then immersed in an electroless nickel bath, and the palladium metal spots catalyze the reduction of the nickel ions to nickel metal by the hypophosphite ions.

Nickel metal deposits first on the palladium spots and then catalyzes the deposition of more nickel. The nickel-covered spots then begin to spread, covering the entire plastic surface with a continuous nickel coating (containing a little phosphorus). This nickel-coated surface is very thin but conducts enough electricity so that the plastic piece can then be plated electrically in a standard electrobath.

In common practice the chemical nickel coating is plated over with a layer of electrocopper, then a layer of electrobright nickel, and finally a layer of chrome. The electrocopper layer fills in the uneven dull surface of the electroless nickel–plastic surface to provide a smooth foundation for subsequent coatings. Also, the ductility of copper acts as a cushion— it absorbs the difference in thermal expansion between the plastic base and the metal coatings and thus prevents peeling. However, since copper is not corrosion resistant, it is protected by a bright nickel coating which is. Why must the nickel coating be bright? Since the final chrome layer is simply decorative, it is so thin that it is transparent. The total thickness of the metal coatings is only about 0.0015 millimeter. Thus, in a chrome-plated plastic piece, the top layer is chrome, below that is bright nickel, then copper, and finally the electroless nickel plate which is anchored to the plastic surface by small dots of palladium anchored to the porous surface of the little pits etched on the surface of the plastic. When properly done, this metal coating adheres tenaciously to the plastic surface. Research is in progress to plate glass, wood, cellophane, steel, and even plastic cloths with electroless nickel as an undercoating.

13

Miscellaneous Industrial Uses for Inorganic Phosphorus Compounds

Polyphosphoric Acid

Polyphosphoric acid is used by petroleum refiners to make cheaper and better gasoline and to make intermediates for detergents and plastics from materials which used to be considered refinery waste and were destroyed. This thick, viscous liquid consists of molecules of different chain lengths. It is produced by heating orthophosphoric acid to drive off its water until a P_2O_5 content of 82–84% is reached. The resulting polyphosphoric acid is used extensively as a catalyst in the petroleum industry for the polymerization, alkylation, dehydrogenation, and isomerization processes described below.

Because of its thickness and high viscosity, polyphosphoric acid is most easily handled when it adheres to and partially reacts with some porous material such as diatomaceous earth. Its catalytic activity, which depends largely on its hydrogen ions, decreases when it is severely dehydrated to metaphosphoric acid (long chain phosphoric acid with only one hydrogen per phosphorus atom).

Since dehydration begins at around 425°F (232°C), most reactions using this catalyst are carried out below this temperature. To suppress dehydration, wherever conditions permit, 2–10% water is usually introduced as steam with the reactants. In the petroleum industry, spent polyphosphoric catalyst is often regenerated by passing air over or through it to burn off the carbonaceous residue. The resulting metaphosphoric acid is then broken down to lower polyphosphoric acids by reaction with steam at about 500°F (260°C).

Polymerization is the process of building large molecules by joining smaller ones into repeating chains. In petroleum refining, large molecules in crude oil are cracked into molecules of different sizes. In the early days of gasoline production much larger quantities of small molecules such as propylene (C_3H_7) were produced than could be sold, and these were disposed of by burning. Later it was found that propylene could be polymerized, using polyphosphoric acid as catalyst, to form olefins (9 or 12 carbons in the chain). These compounds are large enough to be added to the gasoline fraction in concentrations up to several percent.

Polyphosphoric acid catalyst promotes only the linkage between a few propylene units whereas modern polymerization catalysts join thousands of propylene units to form solid plastics known as polypropylene.

An olefin containing 12 carbons can react with a benzene ring to form an alkylbenzene. This alkylation reaction is also catalyzed by polyphosphoric acid. Alkylbenzene, when sulfonated by sulfuric acid, has been used extensively in detergent formulations. Less is used now because of the demand for non-polluting surfactants which are more readily digested by natural plants and microbes. Polymerization and alkylation reactions are shown in the following simplified equations:

Polymerization:

$$4 \; \underset{\underset{\text{H}}{|}}{\overset{\overset{\text{CH}_3}{|}}{\text{C}}} = \text{CH}_2 \quad \xrightarrow{\underset{\text{acid}}{\text{polyphosphoric}}} \quad \text{HC}{=}\text{CH}_2{-}\underset{\underset{\text{H}}{|}}{\overset{\overset{\text{CH}_3}{|}}{\text{C}}}{-}\text{CH}_2{-}\underset{\underset{\text{H}}{|}}{\overset{\overset{\text{CH}_3}{|}}{\text{C}}}{-}\text{CH}_2{-}\underset{\underset{\text{H}}{|}}{\overset{\overset{\text{CH}_3}{|}}{\text{C}}}{-}\text{CH}_3$$

propylene   C-12 olefin

Alkylation:

$$\text{HC}{=}\underset{\underset{\text{H}_2}{}}{}\text{CH}_2{-}\underset{\underset{\text{H}}{|}}{\overset{\overset{\text{CH}_3}{|}}{\text{C}}}{-}\text{CH}_2{-}\underset{\underset{\text{H}}{|}}{\overset{\overset{\text{CH}_3}{|}}{\text{C}}}{-}\text{CH}_2{-}\underset{\underset{\text{H}}{|}}{\overset{\overset{\text{CH}_3}{|}}{\text{C}}}{-}\text{CH}_3 \quad + \quad$$

C-12 olefin   benzene

$$\xrightarrow{\underset{\text{acid}}{\text{polyphosphoric}}} \quad \text{CH}_3{-}\underset{\underset{\text{H}}{|}}{\overset{\overset{\text{CH}_3}{|}}{\text{C}}}{-}\text{CH}_2{-}\underset{\underset{\text{H}}{|}}{\overset{\overset{\text{CH}_3}{|}}{\text{C}}}{-}\text{CH}_2{-}\underset{\underset{\text{H}}{|}}{\overset{\overset{\text{CH}_3}{|}}{\text{C}}}{-}\text{CH}_2{-}\overset{\overset{\text{CH}_3}{|}}{\text{CH}}{-}$$

alkylbenzene

The alkylation of benzene with propylene to form cumene (isopropylbenzene) has become an important industrial process. Cumene is an intermediate in a process which simultaneously produces phenol and acetone—both useful industrial chemicals. The alkylation reaction for cumene formation is:

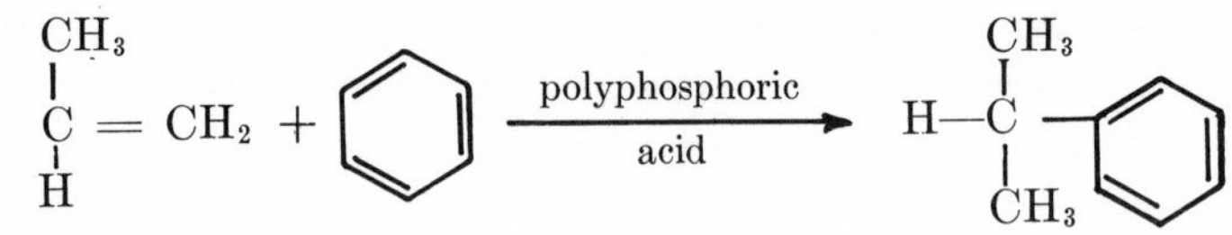

$$\begin{array}{c} CH_3 \\ | \\ C = CH_2 \\ | \\ H \end{array} + \bigcirc \xrightarrow[\text{acid}]{\text{polyphosphoric}} \quad H-\underset{\underset{CH_3}{|}}{\overset{\overset{CH_3}{|}}{C}}-\bigcirc$$

propylene   benzene   cumene or
                                                              isopropylbenzene

Dehydrogenation means the removal of hydrogen from hydrocarbon molecules—*e.g.*, the dehydrogenation of ethane into ethylene:

Dehydrogenation:

$$CH_3-CH_3 \xrightarrow[\text{catalyst}]{\text{polyphosphoric acid}} CH_2{=}CH_2 + H_2$$

  ethane   ethylene   hydrogen

This reaction proceeds at a lower temperature and more efficiently in the presence of polyphosphoric catalyst than at the high temperature and pressure used in cracking. Ethylene is valuable in synthesizing polyethylene and vinyl plastics.

Isomerization converts straight chain hydrocarbons into branched chain compounds. The reaction below is important in the petroleum industry. Of the hydrocarbons suitable for gasoline, those with branched chains have higher octane ratings than those with straight chains. Thus, branched chain hydrocarbons give less knock in an automobile engine.

Isomerization:

$$CH_3CH_2CH_2CH_3 \xrightarrow[\text{acid}]{\text{polyphosphoric}} \begin{array}{c} CH_3 \\ | \\ CH_3-C-CH_3 \\ | \\ H \end{array}$$

Although polyphosphoric acid has been used as a catalyst for many other organic reactions, most have not reached the commercial scale.

Monocalcium Phosphate Monohydrate

Effervescent Tablets. Effervescent tablets containing monocalcium phosphate monohydrate give off bubbles when added to water. This bubbling is the quick release of carbon dioxide gas which is caused by the action of acidic monocalcium phosphate on sodium bicarbonate, a type of baking powder, in tablet form. Children use this rapidly generated gas to power toy rockets. If they use their mother's baking powder and do not replace the container lid tightly, moisture seeps in. When this powder is used to bake a cake, the batter will rise to only half its normal height.

Dipotassium Phosphate

Two moles of potassium hydroxide react with one mole of phosphoric acid to give dipotassium phosphate. The anhydrous form of this compound is very soluble in water (at 30°C about 70 grams will dissolve per 100 grams of water). To obtain the compound in crystalline form, water is evaporated under vacuum at around 60°C.

If the water were boiled off at atmospheric pressure rather than under vacuum—*e.g.*, if the mixture were heated to too high a temperature—two moles of dipotassium phosphate would combine, with the elimination of one mole of water, to form tetrapotassium pyrophosphate as a by-product:

$$
\underset{\substack{\text{dipotassium}\\\text{phosphate}}}{(KO)_2PO H + HO P(OK)_2} \longrightarrow \underset{\substack{\text{tetrapotassium}\\\text{pyrophosphate}}}{(KO)_2POP(OK)_2} + H_2O
$$

Dipotassium phosphate is used as a buffer in car radiators (a buffer is a compound used to maintain the pH of a solution within a desired range). Anti-freeze solutions such as water–glycol mixtures become acidic (lower pH) with time through oxidative reactions. This acid mixture corrodes the radiator metal, and leaks develop. Dipotassium phosphate buffer maintains a pH of around 9 which prevents corrosion.

Tripotassium Phosphate

Tripotassium phosphate is made by neutralizing the three acidic hydrogens in one mole of phosphoric acid with three moles of potassium hydroxide. The solution can be concentrated and then cooled to yield crystals. However, it is much easier to heat it to 300°–400°C and drive off the water quickly. The compound is not affected by high temperature.

Tripotassium phosphate is used to manufacture styrene–butadiene synthetic rubber by polymerization. It acts as an electrolyte to regulate the polymerization rate and controls the stability of the resulting synthetic latex.

Tripotassium phosphate is also used to remove hydrogen sulfide gas (H_2S) from industrial vapors. Since it is a weak acid, H_2S is absorbed in the basic water solution of tripotassium phosphate. When the solution is heated later, H_2S gas is driven off, and the tripotassium phosphate can be used again. This absorption process is called scrubbing; it recovers H_2S as a valuable by-product and prevents its rotten odor from polluting the atmosphere.

Potassium Polymetaphosphate

Potassium polymetaphosphate can be made into a ball which bounces like rubber—an unusual property for an inorganic compound. Since its molecules are packed tightly together in long chains, it is practically insoluble in water. When a sodium salt is added, the sodium replaces some of the potassium and loosens the chain. If enough sodium is added, the insoluble salt can be made into a viscous gel which still consists of long-chain molecules. At this point, if a bivalent ion such as calcium ion is added, each calcium ion can replace two sodium or two potassium ions from two separate chains. The net result is that the chains are tied together by calcium ions. This is called crosslinking.

A compound crosslinked in this way is no longer water soluble. The many small holes in between the network of long chains can hold some water. The finished product is a rubbery gum which can be stretched into a film or squeezed into a round, bouncing ball. Unfortunately, the water cannot be held permanently. It evaporates with time, and when it is dry, the material becomes a solid powder.

Diammonium Phosphate

The propensity of diammonium phosphate to decompose slowly to form ammonia and the acidic monoammonium phosphate makes it useful as a dye-levelling agent in woolen dye baths—that is, it promotes even dyeing. Colloidal wool dyes that have a high affinity for wool yield color that sturdily resists washing and perspiration. However, the dye tends to precipitate too rapidly on the surface of the cloth, giving uneven color distribution and causing streaks. Rapid deposition of the dye also prevents deep penetration of the color into the fabric. If the bath is kept alkaline, the colloidal dye remains dispersed in the bath and will not precipitate on the fabric.

Diammonium phosphate promotes and maintains an alkaline bath. The wool swells and allows the dye to penetrate. When the dye solution is boiled, the diammonium phosphates gradually decompose, evolving ammonia, and the bath gradually becomes acidic. A typical bath may contain from 0.1 to 0.5% diammonium phosphate; the pH may start at 7.85, and after 2.5 hours of heating, drop to a pH of 5.78. As the bath becomes more acidic, the dye starts to precipitate, but since the change in pH is gradual, dye precipitation is also gradual, and even dyeing with good penetration is obtained.

Tetrasodium Pyrophosphate

Tetrasodium pyrophosphate can sequester heavy metal ions such as iron and vanadium, an important factor in bleaching textiles. Many cotton textiles are bleached by peroxy compounds such as hydrogen peroxide, and peroxides are easily decomposed in the presence of even a trace of heavy metal ions. TSPP and sodium silicate are added to sequester these metal ions, preventing them from catalytically decomposing the peroxides.

Sodium Tripolyphosphate as Dispersing Agent

The tripolyphosphate anion in detergents which has many negative charges is an effective dispersing agent not only for dirt but also for many industrially important materials. For example, in drilling an oil well it is

Lil and Al Bloom—Black Star

Oil well outside Canton, Ohio. Sodium tripolyphosphate is used as a liquid lubricant to cool the drill bit as it bites through rock. It is also often used as the dispersion agent to keep the drilling mud in proper suspension.

necessary to use a liquid lubricant to cool the drill bit as it bites through rock. In addition a liquid is needed to suspend and carry the cuttings to the surface. For most oil well operations, a water suspension of clay is used with a high density compound such as barium sulfate. Sodium tripolyphosphate is often used as the dispersion agent to keep the drilling mud in proper suspension.

Another example is found in the manufacture of cement and brick. In "wet process" cement production the raw materials are ground together in the form of a water slurry, and then they are calcined to remove water. It is advantageous to use as little water as possible and yet obtain the desired viscosity in the reactant mixture (as usual, it is less costly to evaporate less water). Addition of a fraction of 1% sodium tripolyphosphate does this very well. It makes it possible to handle more solid per liquid volume, and the production rate of a cement plant can be increased as much as 10%.

Phosphoric Acid

Phosphoric acid is used as a bonding agent for high-alumina refractory products. Such refractories withstand the high temperatures encountered in the steel industry. Phosphoric acid can also be used for bonding silica or magnesia to produce refractory molds used in metal casting.

Phosphine

Some uses for phosphine depend on its toxic properties—for example, as a fumigant against insects and rodents in stored products. For this purpose, phosphine is generated on the spot from aluminum phosphide (AlP). The commercial product is a tablet of aluminum phosphide and ammonium carbamate impregnated with paraffin. Upon reaction with water or atmospheric moisture, phosphine is released. The ammonium carbamate decomposes to carbon dioxide and ammonia, and these gases prevent the phosphine generated from igniting spontaneously.

Zinc phosphide (Zn_3P_2) has long been used as a rodent poison and in grain baits against field mice and gophers. The acute oral toxicity of zinc phosphide to rats is $LD_{50} = 45.7$ mg/kg of body weight. LD_{50} means the minimum lethal dose required to kill 50% of the rats in a given test.

14

Miscellaneous Industrial Uses for Organic Phosphorus Compounds

Phosphorus Chloride Derivatives—Miscellaneous Uses

Although we seldom encounter phosphorus chlorides in our everyday life, they are important chemical agents for making products we need—*e.g.*, mild synthetic bar soaps, clear non-yellowing plastic sheets, compounds for silver polishes and insecticides, and compounds for extracting uranium for the atomic bomb. The common phosphorus chlorides are:

PCl_3	PCl_5	$POCl_3$	$PSCl_3$
phosphorus trichloride	phosphorus pentachloride	phosphorus oxychloride	phosphorus thiochloride

Phosphorus Pentachloride. PCl_5 is a free-flowing solid made by chlorinating elemental phosphorus. Liquid phosphorus trichloride (PCl_3) is first formed as the intermediate and then is converted immediately into the solid pentachloride using excess chlorine:

$$P_4 + 6\ Cl_2 \longrightarrow 4\ PCl_3$$
phosphorus   chlorine   phosphorus
                                            trichloride

$$4\ PCl_3 + 4\ Cl_2 \longrightarrow 4\ PCl_5$$
                                            phosphorus
                                            pentachloride

Phosphorus pentachloride then reacts with moisture in the air to give off hydrogen chloride (HCl) and phosphorus oxychloride ($POCl_3$). PCl_5 is used mainly as a special chlorinating agent in experimental reactions. Its applications are not industrially important. Of the four phosphorus chlorides listed above, only phosphorus trichloride and phosphorus oxychloride are industrially important.

Phosphorus Trichloride Derivatives. To manufacture phosphorus trichloride (PCl_3) industrially, elemental phosphorus is suspended in previously prepared phosphorus trichloride. Chlorine gas is bubbled in to convert the elemental phosphorus into more phosphorus trichloride; this product is then purified by distillation.

Phosphorus trichloride is highly reactive. When mixed with water, it erupts, giving off hydrochloric acid and phosphorous acid:

$$PCl_3 \quad + \quad 3\ H_2O \longrightarrow H_3PO_3 \quad + \quad 3\ HCl$$

phosphorus trichloride $\qquad$ phosphorous acid $\qquad$ hydrochloric acid

The most important use for phosphorus trichloride is as an intermediate or reagent in the manufacture of many industrial chemicals, ranging from insecticides to synthetic surfactants to ingredients for silver polish. For example, PCl_3 is heated with sulfur in the presence of catalysts such as aluminum chloride or activated charcoal to give phosphorus thiochloride ($PSCl_3$). Phosphorus trichloride is also used as an intermediate in synthesizing phosphorus oxychloride.

As an intermediate, phosphorus trichloride is almost unparalleled in its wide ranging applications. It is used to make acyl chlorides from which we get synthetic surfactants, bar soaps, and detergents; it is used to make alkyl chlorides, the intermediate for mercaptans which are com-

Courtesy Stauffer Chemical Co.

PVC resin products find many varied uses—from plastic packaging to the printed woodgrain film on station wagon exteriors. The automotive, construction, furniture, and decorative interior industries are primary markets for these products. Phosphorus compounds are used as PVC stabilizers.

ponents in silver polishes, synthetic rubber, and vinyl plastics; PCl_3 is also used to make dialkyl phosphonates which are themselves used to make insecticides, wetting agents, and metal extractants. Miscellaneous reactions use phopsorus trichloride as an intermediate in nylon manufacture and resin stabilizers. These uses and reactions are discussed in detail below.

PHOSPHOROUS ACID. Phosphorous acid (the reaction product of PCl_3 and water shown in the reaction above) in the form of its lead salt (basic lead phosphite, $2 PbO \cdot PbHPO_3$) is used in formulations to stabilize polyvinyl plastics—*i.e.*, to prevent them from discoloring.

Another commercially important class of compounds derived from phosphorous acid goes by the trade name of Dequest. It is prepared by the reaction of phosphorous acid with formaldehyde and ammonium chloride in the presence of hydrochloric acid:

$$3H_3PO_3 \quad + \quad NH_4Cl \quad + \quad 3CH_2O \quad \overset{HCl}{\underset{H_2O}{\rightarrow}} \quad N[CH_2P(OH)_2]_3$$

phosphorous acid   ammonium chloride   formaldehyde   Dequest aminotri (methyl-phosphonic acid)

Dequest behaves similarly to sodium pyrophosphate, sodium tripolyphosphate, and glassy sodium polyphosphate in that it is an efficient chelating agent for metal ions such as iron, calcium and magnesium. Thus it is useful in detergent compositions for softening water, for cooling water tower systems and for boiler water treatment to prevent scale formation by the threshold effect (effective at 1-10 ppm concentration). It is also an effective corrosion inhibitor for metals when used in a water solution along with zinc chloride. Despite its higher cost, Dequest is used in place of sodium polyphosphates in many of the above applications. It has the advantage of being stable under pH and temperature conditions where polyphosphates would lose their effectiveness by decomposing to orthophosphates.

ACYL AND ALKYL CHLORIDES AND THEIR DERIVATIVES FOR SURFACTANTS AND MERCAPTANS. If you have ever worked with phosphorus trichloride and observed first-hand its violent reactive character and corrosive properties, it is hard to believe that it could be used to make any product which is mild to the skin. However, it is a fact that phosphorus trichloride is an effective chlorinating agent that makes it an ideal intermediate in preparing surfactants for mild, bar-type synthetic detergents. It converts fatty acids into fatty acid chlorides. For example, it is used to convert lauric acid (derived from cocoanut oil) into lauroyl chloride:

$$PCl_3 \; + \; 3\; CH_3(CH_2)_{10}COOH \rightarrow 3\; CH_3(CH_2)_{10}COCl \; + \; H_3PO_3$$

phosphorus   lauric acid   lauroyl chloride   phosphorous
trichloride   acid

These fatty acid chlorides are intermediates for synthetic surfactants. When lauroyl chloride reacts with isethionic acid ($HOCH_2CH_2SO_3H$), lauroyl isethionic acid is formed. The sodium salt of this acid is the actual surfactant used in a popular bar detergent (in contrast to bar soap). When this detergent cleans the skin, it removes only a minimum of the skin's natural oil. Surfactants normally used in laundry detergents, such as sodium alkylbenzenesulfonate, discussed previously, are too efficient as cleansers; they remove everything from the skin, including natural, protective oils, resulting in dry, itchy skin. Sodium lauroyl isethionate is prepared by the following reaction:

$$CH_3(CH_2)_{10}CO\overline{Cl \; + \; H}OCH_2CH_2SO_3H$$

lauroyl chloride   isethionic
                                              acid

$$\xrightarrow{\text{NaOH}} \; CH_3(CH_2)_{10}COOCH_2CH_2SO_3Na$$

sodium lauroyl
isethionate

The fatty acid from tallow—tallow acid—is converted to the tallow acid chloride by phosphorus trichloride. The sodium salt of the reaction product of tallow acid chloride with taurine ($H_2NCH_2CH_2SO_3H$), known as sodium tallow taurate, is also used as the surfactant for another type of gentle synthetic detergent bar. The reaction for preparing sodium tallow taurate is:

$$\text{tallow-}CO\overline{Cl \; + \; H}\overset{H}{N}CH_2CH_2CH_2SO_3H$$

tallow acid   taurine
chloride

$$\xrightarrow{\text{NaOH}} \; \text{tallow-}CONHCH_2CH_2SO_3Na \; + \; NaCl$$

sodium tallow taurate

Phosphorus trichloride also chlorinates alkyl alcohols to their corresponding alkyl chlorides. For example, phosphorus trichloride reacts with octyl alcohol [$CH_3(CH_2)_6CH_2OH$] to give octyl chloride:

$$PCl_3 \;+\; 3\, CH_3(CH_2)_6CH_2OH \rightarrow 3\, CH_3(CH_2)_6CH_2Cl \;+\; H_3PO_3$$

phosphorus   octyl alcohol   octyl chloride   phosphorous
trichloride   acid

One of the most important uses of alkyl chloride is as an intermediate for preparing mercaptans by reaction with sodium hydrosulfide, NaSH, as shown below:

$$CH_3(CH_2)_6CH_2Cl \;+\; NaSH \longrightarrow CH_3(CH_2)_6CH_2SH \;+\; NaCl$$

octyl chloride   sodium   octyl mercaptan   sodium
                    hydrosulfide   chloride

Octyl mercaptan is used in silver polish. It reacts with tarnish (silver sulfide) to form silver octyl mercaptide, a colorless product which is easily wiped off. It also leaves a very thin coating of silver octyl mercaptide on the surface of the silver. This coating protects the silver from the tarnishing action of hydrogen sulfide in the air.

Long-chain alkyl mercaptans are also used to control molecular weight in synthetic rubber manufacture. Some tin organic mercaptides are used as stabilizers for vinyl plastics.

DIALKYL PHOSPHONATES AS INTERMEDIATES FOR INSECTICIDES, WETTING AGENTS, AND METAL EXTRACTANTS. Phosphorus trichloride is the intermediate used to prepare the following important classes of com-

pounds: the dialkyl phosphonates $[(RO)_2\overset{\text{O}}{\overset{\|}{P}}H]$, the trialkyl phosphites $[(RO)_3P]$, and triaryl phosphites $[(ArO)_3P]$. (R represents an alkyl group, and Ar refers to an aryl group.)

The use of phosphorus trichloride as an intermediate in preparing dialkyl phosphonates, as exemplified by diisopropyl phosphonate $[(i\text{-}C_3H_7O)_2\overset{\text{O}}{\overset{\|}{P}}H]$, for the nerve gas DFP, is discussed in Chapter 17. The series of reactions, starting with phosphorus trichloride and ending with DFP, is summarized in the following three equations:

(1) $PCl_3 \;+\; 3\, i\text{-}C_3H_7OH \rightarrow (i\text{-}C_3H_7O)_2\overset{\text{O}}{\overset{\|}{P}}H \;+\; i\text{-}C_3H_7Cl \;+\; 2\, HCl$

       phosphorus   isopropyl   diisopropyl   isopropyl   hydrogen
       trichloride   alcohol   phosphonate   chloride   chloride

(2) $(i\text{-}C_3H_7O)_2\overset{\displaystyle O}{\overset{\|}{P}}H$ + Cl$_2$ → $(i\text{-}C_3H_7O_2)\overset{\displaystyle O}{\overset{\|}{P}}Cl$ + HCl
                              chlorine   diisopropyl
                                             phosphorochloridate

(3) $(i\text{-}C_3H_7O)_2\overset{\displaystyle O}{\overset{\|}{P}}Cl$ + NaF   → $(i\text{-}C_3H_7O)_2\overset{\displaystyle O}{\overset{\|}{P}}F$ + NaCl
                              sodium   diisopropyl
                              fluoride   phosphorofluoridate
                                             (DFP)

Reaction 1 is not limited to isopropyl alcohol. Other alkyl alcohols react similarly with phosphorus trichloride to give the corresponding dialkyl phosphonate $[(RO)_2\overset{\displaystyle O}{\overset{\|}{P}}H]$. For example, with methyl alcohol (CH_3OH), the product is dimethyl phosphonate:

PCl$_3$ + 3 CH$_3$OH ⟶ $(CH_3O)_2\overset{\displaystyle O}{\overset{\|}{P}}H$ + CH$_3$Cl + 2 HCl
          methyl   dimethyl   methyl   hydrogen
          alcohol   phosphonate   chloride chloride

With octyl alcohol, $C_8H_{17}OH$, the product is dioctyl phosphonate:

PCl$_3$ + 3 C$_8$H$_{17}$OH ⟶ $(C_8H_{17}O)_2\overset{\displaystyle O}{\overset{\|}{P}}H$ + C$_8$H$_{17}$Cl + 2 HCl
          octyl   dioctyl   octyl
          alcohol   phosphonate   chloride

Dimethyl phosphonate from the reaction of phosphorus trichloride with methyl alcohol is the intermediate used in making the insecticide Dipterex:

$(CH_3O)_2\overset{\displaystyle O}{\overset{\|}{P}}H$ + Cl$_3$CCHO ⟶ $(CH_3O)_2\overset{\displaystyle O}{\overset{\|}{P}}CHCCl_3$
dimethyl   chloral   O
phosphonate   H

                                            Dipterex
                                            (dimethyl 1-hydroxy-
                                            2,2,2-trichloroethylphosphonate)

Dioctyl phosphonate [(C$_8$H$_{17}$O)$_2$$\overset{\displaystyle O}{\overset{\|}{P}}$H], obtained by the action of phosphorus trichloride on octyl alcohol, can be chlorinated to give the dioctyl

phosphorochloridate [(C$_8$H$_{17}$O)$_2$$\overset{\displaystyle O}{\overset{\|}{P}}$Cl], which upon hydrolysis gives dioctyl

phosphoric acid [(C$_8$H$_{17}$O)$_2$$\overset{\displaystyle O}{\overset{\|}{P}}$OH]:

$$(C_8H_{17}O)_2\overset{O}{\overset{\|}{P}}H \ + \ Cl_2 \longrightarrow (C_8H_{17}O)_2\overset{O}{\overset{\|}{P}}Cl \ + \ HCl$$

dioctyl phosphonate   chlorine   dioctyl phosphoro-chloridate

$$(C_8H_{17}O)_2\overset{O}{\overset{\|}{P}}Cl \ + \ H_2O \longrightarrow (C_8H_{17}O)_2\overset{O}{\overset{\|}{P}}OH \ + \ HCl$$

dioctyl phosphoric acid

The sodium salt of dioctyl phosphoric acid is a very good wetting agent and is used to some extent for this purpose. Dioctyl phosphoric acid, in which the octyl alcohol is 2-ethylhexyl alcohol (the product is then properly called di-2-ethylhexyl phosphoric acid), is important in extracting uranium from its ore. The ore is dissolved in 10% sulfuric acid, and the uranium solution is then extracted with kerosene containing 5–6% di-2-ethylhexyl phosphoric acid. Other impurities are left in the original sulfuric acid solution, and uranium is extracted from the kerosene as a fairly pure sodium salt using a 10% solution of soda ash.

TRIALKYL AND TRIARYL PHOSPHITES FOR INSECTICIDES AND POLYMER STABILIZERS. Trialkyl phosphites are prepared by the reaction of phosphorus trichloride with alcohols in the presence of a base, such as a tertiary amine (R$_3$N), which absorbs the hydrogen chloride (HCl) evolved. If an absorbing agent were not used, HCl would attack the trialkyl phosphite formed and convert it into a dialkyl phosphonate. The general reaction for preparing trialkyl phosphites is:

$$PCl_3 \ + \ 3\ ROH \ + \ 3\ R_3N \ \rightarrow \ (RO)_3P \ + \ 3\ R_3N \cdot HCl$$

phosphorus trichloride   alkyl alcohol   tertiary amine   trialkyl phosphite   tertiary amine hydrochloride

If methyl alcohol is used, trimethyl phosphite $[(CH_3O)_3P]$, is formed. This is probably the most important trialkyl phosphite produced today. Its main use is as the intermediate in synthesizing a series of insecticides such as DDVP and Phosdrin. For example, the reaction to make DDVP involves the action of trimethyl phosphite on chloral (Cl_3CCHO) (*see* Chapter 19).

Other trialkyl phosphites are used industrially. For example, triiso-octyl phosphite $[(i\text{-}C_8H_{17}O)_3P]$, which is prepared from isooctyl alcohol ($i\text{-}C_8H_{17}OH$), is a stabilizer for polyvinyl chloride plastics. When vinyl plastic is overheated or exposed to ultraviolet radiation, it discolors. One theory is that the action of heat and ultraviolet radiation causes polyvinyl chloride plastics to lose hydrogen chloride (HCl) and absorb oxygen. To prevent this degradation, selected trialkyl phosphites may be added. Phosphite has a trivalent phosphorus atom which can react with oxygen to form the stable pentavalent phosphate. It thus consumes the oxygen and prevents it from attacking the polymer. The organic groups in the phosphite make it soluble and compatible with the polymer system.

Triphenyl phosphite is made from phenol and PCl_3:

$$PCl_3 \quad + \quad 3\ C_6H_5OH \longrightarrow (C_6H_5O)_3P \quad + \quad 3\ HCl$$

| phosphorus trichloride | phenol | | triphenyl phosphite | |

In the above reaction phosphorus trichloride reacts directly with phenol; no base is required (as it is for synthesizing trialkyl phosphites) because the hydrochloric acid evolved does not attack triaryl phosphite $(ArO)_3P$. In the case of trialkyl phosphite $[(RO)_3P]$, if a base is not used to absorb the acid, HCl will convert trialkyl phosphite to dialkyl phosphonate as shown below:

$$
(RO)_3P \quad + \quad HCl \longrightarrow (RO)_2\overset{\displaystyle O}{\overset{\displaystyle \|}{P}}H \quad + \quad RCl
$$

| trialkyl phosphite | hydrogen chloride | | dialkyl phosphonate | alkyl chloride |

Obviously it is easier to make triphenyl phosphite than trialkyl phosphite. In fact, chemists often prefer to make triphenyl phosphite first and then convert it to the trialkyl phosphite by ester exchange. The following reaction illustrates the preparation of tridecyl phosphite by the ester exchange reaction; here decyl alcohol replaces phenol:

$$(C_6H_5O)_3P + 3\ C_{10}H_{21}OH \longrightarrow (C_{10}H_{21}O)_3P + 3\ C_6H_5OH$$

| triphenyl phosphite | decyl alcohol | | tridecyl phosphite | phenol |

Phenol can be recovered for re-use with phosphorus trichloride to make more triphenyl phosphite to start the reaction over again.

Triphenyl phosphite is used alone as a resin stabilizer—*e.g.*, in cooking alkyd resins used in enamel coatings for large appliances. A small amount of it in the reaction mixture results in a lighter-colored resin. Triphenyl phosphite is also used in stabilizer formulations for vinyl plastics. A special triaryl phosphite, tri(nonylphenyl) phosphite $[(C_9H_{19}C_6H_4O)_3P]$, is a very effective stabilizer for the GRS (government rubber–styrene, made during World War II) type of synthetic rubber and polyolefins. It protects them from degradation by heat and light.

PHENYLPHOSPHONOUS DICHLORIDE AND DERIVATIVES USED IN NYLON AND INSECTICIDES. Phosphorus trichloride reacts at 600°–700°C with benzene (C_6H_6) to give phenylphosphonous dichloride:

$$\underset{\substack{\text{phosphorus} \\ \text{trichloride}}}{PCl_3} \quad + \quad \underset{\text{benzene}}{C_6H_6} \quad \xrightarrow{\text{600°–700° C}} \quad \underset{\substack{\text{phenylphos-} \\ \text{phonous} \\ \text{dichloride}}}{C_6H_5PCl_2} + HCl$$

Phenylphosphonous chloride ($C_6H_5PCl_2$) reacts with water to give phenyl-

phosphinic acid, $C_6H_5P\overset{\displaystyle O}{\underset{\displaystyle OH}{}}\diagup^{H}$. This acid has been used as stabilizer to

Operator at Monsanto's Blacksburg, S.C., plant watches beaming machine gather nylon fiber for subsequent shipment to textile mill

prevent nylon from discoloring during process heating. Phenylphosphonous dichloride also reacts with sulfur to give phenylphosphonothionic di-

$$\overset{\text{S}}{\overset{\|}{\text{chloride}}}$$

chloride ($C_6H_5PCl_2$). This compound is the intermediate in preparing the
insecticide EPN (Chapter 19).

Phosphorus Oxychloride Derivatives. Phosphorus oxychloride is a
versatile intermediate for manufacturing many important organic phosphates. It is prepared by one of two commercial methods. One method
involves the direct oxidation of phosphorus trichloride with oxygen:

$$2 \text{ PCl}_3 \quad + \quad \text{O}_2 \quad \longrightarrow \quad 2 \text{ POCl}_3$$

phosphorus	oxygen	phosphorus
trichloride		oxychloride

The other method is the chlorination of phosphorus trichloride, PCl_3, in
the presence of phosphoric anhydride (P_4O_{10}).

$$6 \text{ PCl}_3 + \text{P}_4\text{O}_{10} + 6 \text{ Cl}_2 \longrightarrow 10 \text{ POCl}_3$$

Phosphorus oxychloride's most important use is as a reactant in the
manufacture of the triaryl and trialkyl phosphates, as well as the mixed
alkyl aryl phosphates. The discussion of the chemistry and application
of phosphorus oxychloride in this section, therefore, concerns the chemistry and application of the phosphate esters.

TRIARYL PHOSPHATES AND THEIR USES. A typical phosphate, tricresyl
phosphate, is made by heating phosphorus oxychloride with cresol:

$$\text{POCl}_3 \quad + \quad 3 \text{ CH}_3\text{C}_6\text{H}_4\text{OH} \quad \longrightarrow \quad \text{OP(OC}_6\text{H}_4\text{CH}_3)_3 + 3 \text{ HCl}$$

phosphorus	cresol	tricresyl	hydrogen
oxychloride		phosphate	chloride

The first important phosphate esters introduced commercially were
triphenyl and tricresyl phosphates; they were used to make a better
quality "celluloid." Many years ago billiard balls were made of ivory,
but the sport became so popular, an ivory shortage resulted. The first
substitute for ivory was celluloid, a partially nitrated cellulose plasticized
with camphor. Celluloid was also found to be excellent for motion
picture film and toys.

Since the partially nitrated cellulose was made by the controlled
reaction of cellulose (*e.g.*, cotton) with nitric acid in the presence of
sulfuric acid, it was a close relative of gun cotton, a more fully nitrated
cellulose. When blended with camphor as the plasticizer, it was extremely flammable. [A plasticizer is an additive incorporated in the

plastic to soften it and to increase its flexibility and workability.] Movie films and some toys made with it practically exploded when ignited accidentally. Since phosphorus compounds are flame retardants, triphenyl and tricresyl phosphate were checked and found to be effective replacements for camphor as plasticizers. These phosphates do reduce the flammability of cellulose nitrate, but nothing can make cellulose nitrate really flameproof. Eventually, for movie films, cellulose nitrate was replaced with the much slower burning cellulose acetate.

The discovery of triaryl phosphates as plasticizers for cellulose nitrate opened an area for their application as plasticizers for other polymers. They perform very well in cellulose acetate and for vinyl polymers. The use of the triaryl phosphate esters as plasticizers not only permits plastics to be processed at a lower temperature and with less effort but also imparts some flame resistance to the plastics in which they are used.

Aryl phosphates are also used as gasoline additives. When gasoline containing tetraethyl lead burns in an automobile engine, lead deposits form on the spark plugs and on the inside of the cylinder head. These deposits catalytically ignite the gasoline–air mixture in the cylinder prior to ignition by the spark plugs. These misfires are heard as wild pings and reduce gasoline mileage.

Aryl phosphates such as tricresyl phosphate $[(CH_3C_6H_4O)_3P{=}O]$, or cresyl diphenyl phosphates $[(CH_3C_6H_4O)(C_6H_5O)_2P{=}O]$, when added in small percentages to the gasoline, combat misfires. The theory is that when triaryl phosphates are burned in the gasoline mixture, the phosphorus moiety combines with the lead decomposition product to form a phosphorus-containing lead compound, such as lead phosphate. This compound deposits on the spark plug and inside the cylinder heads, and unlike other lead deposits, it is no longer active catalytically—*i.e.*, it does not cause premature firing. According to recent studies, the unleaded gasolines so favored by environmentalists cause erosion in the exhaust valves of some engines. This phenomenon is minimized by phosphorus-containing gasoline additives. This protective action could be the result of the formation of phosphorus-containing deposits substituting for the lead-containing deposits from leaded gasolines.

Because triaryl phosphates are less flammable than petroleum oil, they have gradually replaced the latter as lubricants, coolants, and hydraulic fluids in machinery where fire is a potential danger. For example, triaryl phosphates are used as lubricant and coolant in the large bearings in huge gas turbines which generate electricity and steam. These turbines are also used to pump oil and gas through large pipelines and to drive air compressors in the chemical processing industries.

Triaryl phosphates are used as hydraulic fluids to control the opening and shutting of valves in the electrohydraulic control units of the large

Tributyl phosphate is used as a non-flammable component in the fluid used in the hydraulic system of large commercial aircraft such as the one shown here

equipment. The big machinery in steel mills which control the size, shape, and thickness of steel sheets and bars during processing are also controlled hydraulically. Since the steel pieces are very hot, less flammable fluids must be used in case of accidental leakage in the hydraulic system.

Mining equipment is also hydraulically controlled. Since the threat of fire and explosion inside mines is also present, less flammable fluids such as these based on triaryl phosphate are recommended.

A minor use for tricresyl phosphate is an an additive in lubricating oils to reduce wear on bearing surfaces. Theoretically, under actual driving conditions, tricresyl phosphate gradually decomposes to cresyl acid phosphates that react with the iron surfaces of the bearings, forming a microscopic protective coating over them.

TRIALKYL AND ALKYL ARYL PHOSPHATES AND THEIR USES. Other important phosphate esters prepared from phosphorus oxychloride are the trialkyl and mixed alkyl aryl phosphates. The trialkyl phosphates, as exemplified by tri-2-ethylhexyl phosphate and tributyl phosphate, are prepared by the reaction of phosphorus oxychloride with the respective alcohols. The conditions for making these phosphates are much milder than those used for triaryl phosphates. Tributyl phosphate is prepared in the following way:

$$POCl_3 \; + \; 3 \; C_4H_9OH \; \longrightarrow \; OP(OC_4H_9)_3 + 3 \; HCl$$

| phosphorus | butyl | tributyl |
| oxychloride | alcohol | phosphate |

When alcohols other than butyl are used, the corresponding trialkyl phosphates can be prepared by this general reaction.

Tributyl phosphate is used to purify uranium. As noted before, dioctyl phosphoric acid is used in the solvent extraction of uranium from its ore. Although the uranium compound obtained is fairly pure, it is not pure enough for the conversion to uranium metal for atomic reactors. In one purification process the extracted uranium is dissolved in nitric acid. This solution is extracted further with tributyl phosphate in a kerosene solution. In this step, the tributyl phosphate actually forms a compound with the uranyl nitrate that is soluble in the kerosene solution: $UO_2(NO_3)_2[(C_4H_9O)_3P{=}O]_2$. The uranium salt is then extracted from the kerosene solution with a 10% solution of sodium carbonate. The selectivity of tributyl phosphate in extracting one metal from the other also makes it useful for separating normally difficult-to-separate elements such as hafnium from zirconium.

Tributyl phosphate is also used as a non-flammable component in the fluid used in the hydraulic system of large commercial aircraft. Another important application is as a defoamer in preparing latex paints

Courtesy Stauffer Chemical Co.

The interior and exterior automotive vinyls have been color coordinated in this automobile. The exterior roof, seat inserts, and a strip on the door panels and quarter panels of this car are color matched and finished by a special coating technique developed by the Stauffer Chemical Co. Phosphorous compounds are used to manufacture and stabilize vinyl plastics like these.

and printing inks. Tributyl phosphate is also used in processing industries such as paper manufacturing, where agitation creates foam and results in spillage. Less than 1% of this compound depresses foaming by changing the surface tension of the liquid.

Another important trialkyl phosphate is tri-2-ethylhexyl phosphate. As a plasticizer for vinyl polymer sheets, it not only imparts flame resistance to the sheets, but the plasticized sheets remain flexible at low temperatures. When such sheets are used as seat covers in automobiles and trucks, they will not stiffen or crack in cold weather.

Of the mixed alkyl aryl phosphates, methyl dicresyl phosphate

$$
\overset{\displaystyle O}{\underset{\displaystyle \|}{}}
$$

$[CH_3OP(OC_6H_4CH_3)_2]$, is also used as a gasoline additive. The most important mixed ester, however, is the 2-ethylhexyl diphenyl phosphate

$$
\overset{\displaystyle O}{\underset{\displaystyle \|}{}}
$$

$C_8H_{17}OP(OC_6H_5)_2$. This nontoxic compound finds use as a plasticizer for food wrapping films, films for tubings for skinless sausages, and for other meat packaging. It has been approved for this use by both the Bureau of Animal Industry and the Food and Drug Administration. 2-Ethylhexyl diphenyl phosphate is also the main component in certain nonflammable hydraulic fluids used in large aircraft. Since it does not ignite spontaneously below 1000°F and remains fluid at low temperatures, it is ideal for this purpose.

TRIALKYL TRITHIOPHOSPHATE. A very interesting compound is the reaction product of phosphorus oxychloride with butyl mercaptan (essence of skunk):

$$POCl_3 + 3\ CH_3(CH_2)_2CH_2SH \rightarrow OP[SCH_2(CH_2)_2CH_3]_3 + 3\ HCl$$

phosphorus	butyl	S,S,S-tributyl
oxychloride	mercaptan	trithiophosphate

S,S,S-Tributyl trithiophosphate is a good defoliant; it removes leaves from cotton plants without killing them. When these plants are machine harvested, the green leaves normally stain the white cotton. If the plants are defoliated by a spray of a water emulsion of a compound such as S,S,S-tributyl trithiophosphate, the staining problem is eliminated.

Miscellaneous Uses of Phosphoric Anhydride Derivatives

Phosphoric Anhydride Reactions with Water. Phosphoric anhydride, P_4O_{10}, is generally referred to as P_2O_5. In discussing the chemistry of this compound, it is easier to use the correct formula of P_4O_{10}, with its four-

The effects of defoliation are shown in this aerial photo of road-way (left) and powerline (lower and upper right). S,S,S-Tributyl trithiophosphate is used as a defoliant in harvesting cotton plants.

sided (tetrahedral) structure. Phosphoric anhydride forms when elemental phophorus burns in air. Most P_4O_{10} manufactured is immediately converted into phosphoric acid by reaction with water. Phosphoric anhydride, however, is the water-free form of phosphoric acid. [When phosphorus burns in a limited amount of air, the lower oxide of phosphorus, phosphorous anhydride (P_4O_6) also forms. Since this oxide is not important commercially, it is not discussed in this chapter.]

As a free-flowing white powder, phosphoric anhydride has a strong affinity for water and will readily absorb moisture from the air or any place it can obtain it. This property makes it a natural candidate as a desiccant, and it is widely used for drying small amounts of compounds in chemical laboratories. As P_4O_{10} absorbs water, it becomes a gummy form of polyphosphoric acid. To prevent this gummy material from coating unreacted P_4O_{10}, and thus reducing its dehydrating capacity, P_4O_{10} is generally used as a porous mixture with activated carbon.

Pictorially, the tetrahedral structure of phosphoric anhydride resembles elemental phosphorus with oxygen added. Phosphoric anhydride with its added oxygens still forms a pyramidal structure with three sides and a base. When water reaches P_4O_{10}, it splits the POP bonds—*i.e.*, one mole of water splits one POP bond. Since there are six POP bonds in each P_4O_{10} tetrahedron, six moles of water are required to break all of the POP bonds, resulting in four moles of phosphoric acid [$(HO)_3P{=}O$ or H_3PO_4].

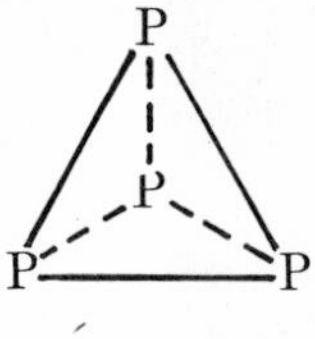

P_4
phosphorus

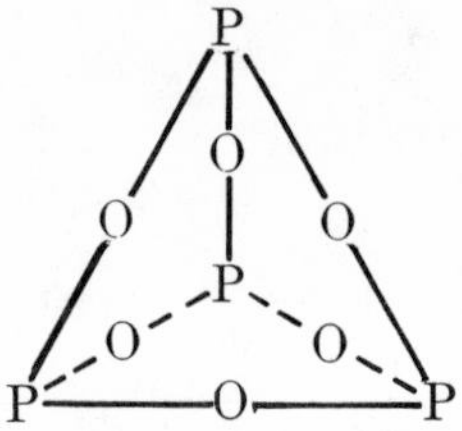

P_4O_6
phosphorous
anhydride

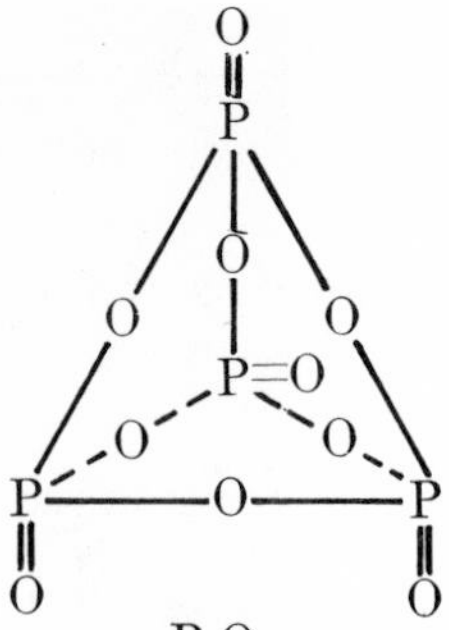

P_4O_{10}
phosphoric
anhydride

By selectively breaking only two bonds with two moles of water (shown below), cyclic tetrametaphosphoric acid is formed:

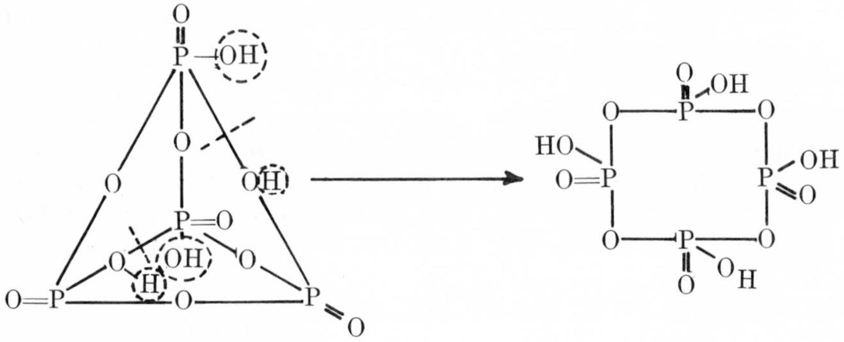

breaking of two POP
bonds with two H_2O

cyclic tetrametaphosphoric
acid

As a matter of fact, one method for preparing sodium cyclic tetrametaphosphate in fairly good yield is by adding P_4O_{10} to a water solution of sodium hydroxide at 0°C. The product can be separated from other sodium phosphates by fractional crystallization.

When water reacts with P_4O_{10}, each POP bond splits into two POH groups. Above 150°C, the two POH groups from two moles of phosphoric acid $[O{=}P(OH)_3]$ can combine and eliminate one mole of water to form a new POP bond:

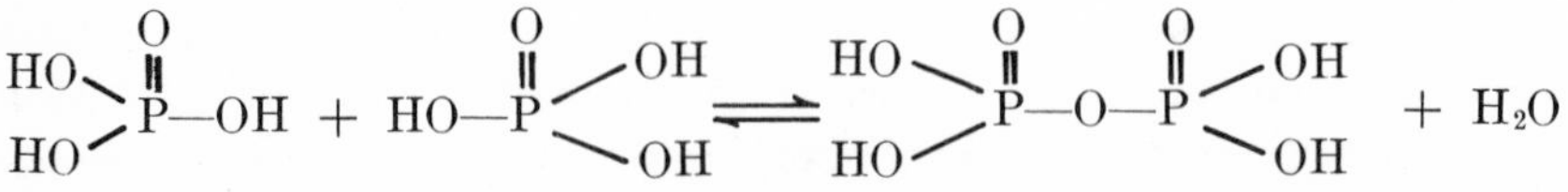

In other words, in the proper temperature range, the splitting of a POP bond by a water molecule and the formation of a new POP bond with the elimination of a mole of water is a reversible, equilibrium reaction.

POH groups can also split another POP and at the same time create a new POP bond. This is illustrated below by the reaction of orthophosphoric acid [$(HO)_3P{=}O$] with tripolyphosphoric acid to form two moles of pyrophosphoric acid.

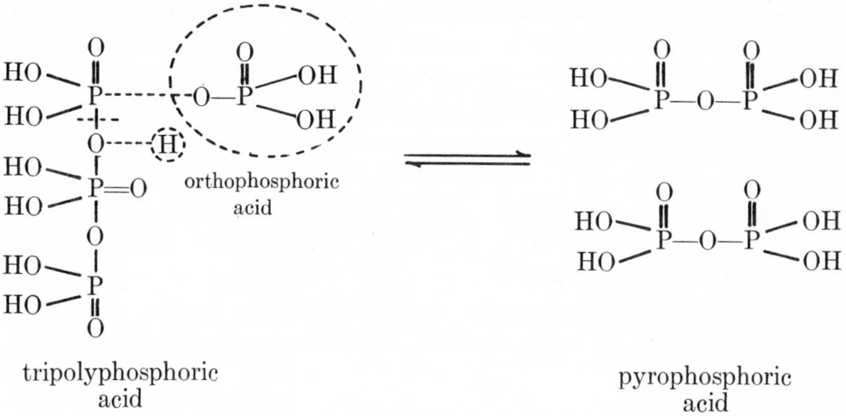

tripolyphosphoric
acid

orthophosphoric
acid

pyrophosphoric
acid

Bond-splitting and new bond formation continue until equilibrium is reached. At that point the product is a mixture of phosphoric acids of various chain lengths. As indicated in the chart in Chapter 3, the composition at equilibrium depends on the concentration of P_4O_{10} in the mixture.

Alkyl Acid Phosphates and Their Uses. The P_4O_{10} tetrahedron also reacts with alcohols, ROH (and phenols, ArOH); the reaction is generally done at about 60°–70°C; ROH reacts with the POP bond as follows:

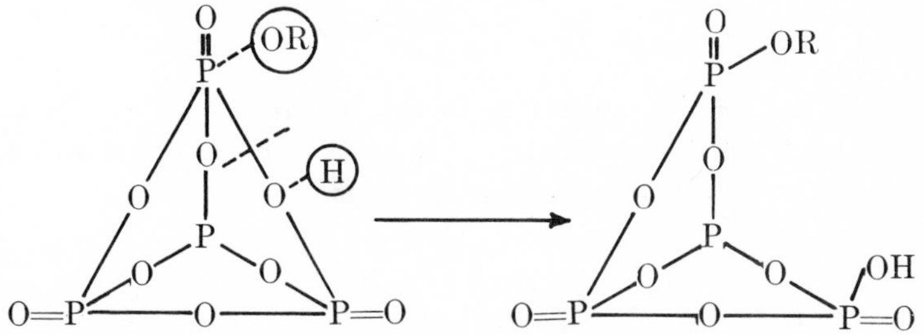

The reaction of ROH with the various POP bonds in the tetrahedron follows the laws of probability; sometimes the OR groups go to the P in one end of the POP bond which already has an RO–group, and sometimes they go to the end of the POP bond without an RO–group. When one mole of P_4O_{10} reacts completely with six moles of ROH, the resulting product is essentially an equimolar (one mole of each) mixture of

$$(RO)_2\overset{\overset{O}{\|}}{P}OH \text{ and } (RO)\overset{\overset{O}{\|}}{P}(OH)_2.$$

The combination of $(RO)_2\overset{\overset{O}{\|}}{P}OH$ and $(RO)\overset{\overset{O}{\|}}{P}(OH)_2$ is called mixed mono- and dialkyl phosphoric acids. The nature of R depends on the alcohol used in the reaction. When ethyl alcohol (C_2H_5OH) is used, the product is a mixture of the mono- and diethyl phosphoric acids

$$[C_2H_5O\overset{\overset{O}{\|}}{P}(OH)_2 \text{ and } (C_2H_5O)_2\overset{\overset{O}{\|}}{P}OH].$$ The ammonium salt of the mono- and diethyl phosphoric acid is very water soluble. It has been promoted as a textile lubricant, making it easier to feed fabric through the machinery; it is also a flameproofing agent. When the alcohol used in the reaction with P_4O_{10} is polyoxyethylenated nonylphenol,

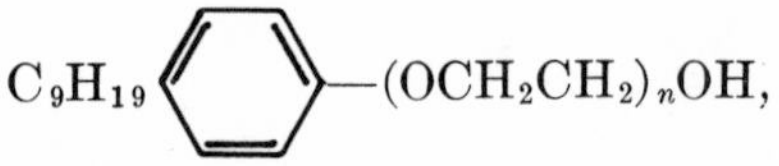

$$C_9H_{19}\text{—}\bigcirc\text{—}(OCH_2CH_2)_nOH,$$

the reaction product is a mixture of

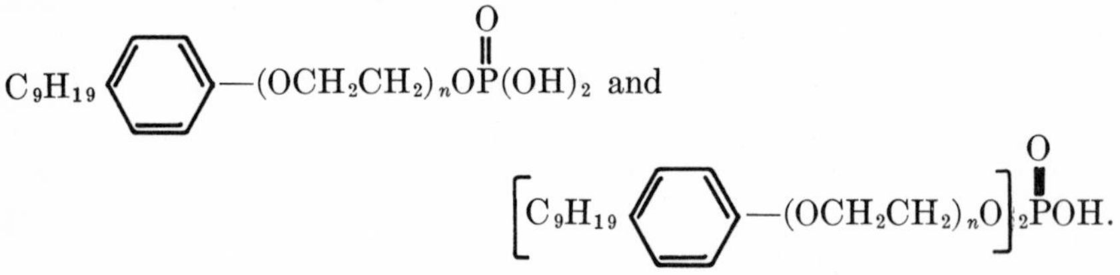

$$C_9H_{19}\text{—}\bigcirc\text{—}(OCH_2CH_2)_nO\overset{\overset{O}{\|}}{P}(OH)_2 \text{ and}$$

$$\left[C_9H_{19}\text{—}\bigcirc\text{—}(OCH_2CH_2)_nO\right]_2\overset{\overset{O}{\|}}{P}OH.$$

The sodium salt of this mixture is soluble in organic solvents and is a good surfactant for dry cleaning—*i.e.*, it helps to remove dirt during dry cleaning just as soap and detergents do during laundering.

The $(RO)_2\overset{\overset{O}{\|}}{P}OH$ and $RO\overset{\overset{O}{\|}}{P}(OH)_2$ mixture is acidic. Since it contains organic groups, it is soluble in organic systems. Many of these acids are used as catalysts in such polymerization reactions as the hardening of urea–formaldehyde resin. When mono- and dibutyl acid phosphates are used as the catalyst (about 3% concentration), they accelerate the hardening of the resin even at low (room) temperatures. Similar catalyst systems are also effective for curing or hardening melamine–formaldehyde resins.

The amine salt of the mono- and diisooctyl phosphoric acid is an effective corrosion inhibitor in gasoline pipes in refineries. An amine salt of mono- and ditridecyl phosphoric acid, when added to gasoline, acts as an anti-stalling agent for automobiles by preventing carburetor icing in cold weather. Most of the other specific applications for mono- and dialkyl phosphoric acids and their salts depend on the nature of the alkyl group as well as the nature of the metal salt.

One interesting compound is prepared by neutralizing medium-length chains (with eight to 12 carbons) of mixed mono- and dialkyl phosphoric acid with ethylene oxide. The product is a good emulsifier and wetting agent that foams only slightly—an important property in dye baths where too much foam causes overflow. The compound is also an effective emulsifier for polymerizing vinyl acetate and for copolymerizing vinyl chloride with vinyl acetate. Since it is also quite stable in acids, it is used as a wetting agent in acid metal cleaning systems.

When P_4O_{10} reacts with less than six moles of alcohol, the product contains unbroken POP bonds. In this respect, it can be regarded as an organic polyphosphate. The sodium salts of the reaction products of three moles of P_4O_{10} with 20 moles of octyl alcohol ($C_8H_{17}OH$) are good wetting agents and effective stabilizers against the degradation of vinyl plastics by heat. Since this type of compound also sequesters heavy metals, it makes a good stabilizer in peracetic acid manufacture. Sequestering action ties up heavy metal impurities, thus preventing their catalytic decomposition of the peracetic acid.

Miscellaneous Uses of Organic Compounds from Phosphorus Sulfides

Classes of Phosphorus Sulfides. Phosphorus sulfides are the products of elemental phosphorus and sulfur, and most give off a rotten-egg odor. Although these materials are malodorous, they are essential intermediates for insecticides to keep our flower gardens beautiful, and to help farmers save their cotton crops for our clothing rather than as food for boll weevils. Also, phosphorus sulfides are ingredients for additives to automobile engine lubricants. They are also used by miners to extract minerals from sand and clay mixtures. The better-known phosphorus sulfides are shown on p. 162. These structures indicate that phosphorus sulfides are formed by inserting sulfur atoms into the phosphorus, P_4, tetrahedron. Phosphorus pentasulfide (actually P_4S_{10}) is the result of the breaking of all the phosphorus to phosphorus (—P—P—) bonds in the P_4 tetrahedron along with the oxidation of all of the phosphorus atoms from valence of zero to a valence of 5. The other phosphorus sulfides can be thought of as the

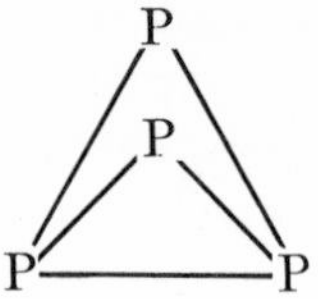

P_4
Phosphorus

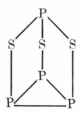

P_4S_3
phosphorus
sesquisulfide

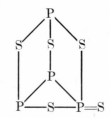

P_4S_5
tetraphosphorus
pentasulfide

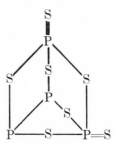

P_4S_7
tetraphosphorus
heptasulfide

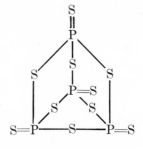

P_4S_{10}
phosphorus
pentasulfide

result of the controlled but incomplete reaction of sulfur with the (P_4) phosphorus tetrahedron. In fact, in manufacturing phosphorus sulfides, the reaction conditions are more or less the same. Except for tetraphosphorus pentasulfide, which is prepared by the reaction of P_4S_3 with sulfur, the other phosphorus sulfides are prepared by heating phosphorus with sulfur at 300°–400°C. The controlling factor in the reaction is the ratio of the sulfur to phosphorus.

In general, stoichiometric quantities of the two elements as represented by the specific formulas are heated together to obtain the desired compound. Depending on the particular phosphorus sulfide obtained, it can be purified by recrystallization from solvents such as carbon disulfide or by distillation. The commercially important phosphorus sulfides are phosphorus sesquisulfide (P_4S_3) (used in safety matches) and phosphorus pentasulfide (P_4S_{10}).

Organic Compounds from Phosphorus Pentasulfide and Their Uses. Phosphorus pentasulfide (P_4S_{10}) is a light-yellow crystalline solid. It reacts rapidly with atmospheric moisture to liberate hydrogen sulfide gas. Its characteristic odor is thus that of rotten eggs, similar to hydrogen sulfide. This unsavory odor carries over to its derivatives. I have worked with many of these derivatives, and for most of them if you say that they smell like a skunk, you are insulting the skunk.

Despite their bad odor, these derivatives have properties which make them valuable in many applications. For example, the insecticide malathion is one of the derivatives of phosphorus pentasulfide.

Phosphorus pentasulfide is a widely used intermediate for synthesizing insecticides, lubricating oil additives, and flotation agents. In all these applications, it reacts with an alcohol to form the dialkyl phosphorodithioic acid, $(RO)_2\overset{\displaystyle S}{\overset{\displaystyle \|}{P}}SH$:

$$P_4S_{10} + 8\ ROH \longrightarrow 4\ (RO)_2\overset{\displaystyle S}{\overset{\displaystyle \|}{P}}SH + 2\ H_2S$$

phosphorus pentasulfide   alcohol   dialkyl phosphorodithioic acid   hydrogen sulfide

The alkyl alcohol (ROH) can be methyl (CH_3OH), ethyl (C_2H_5OH), propyl (C_3H_7OH), or a longer chain alcohol such as octyl ($CH_3(CH_2)_6$-CH_2OH). The alcohol used determines the final application of the product. The low molecular weight alkyl alcohols (methyl, ethyl, and propyl) are generally used to prepare insecticides.

INSECTICIDES. Insecticides such as malathion, Thimet, and Trithion are synthesized as shown on p. 164, starting from phosphorus pentasulfide. Here phosphorus pentasulfide reacts with methyl alcohol to form dimethyl phosphorodithioic acid $[(CH_3O)_2\overset{\displaystyle S}{\overset{\displaystyle \|}{P}}SH]$ which is added to ethyl maleate to produce malathion. This insecticide is widely used to control garden and agricultural pests. In home gardening, therefore the difference between having roses as a thing of beauty to enjoy or as food for insects is the use of an insecticidal dust containing malathion.

Malathion:

$$P_4S_{10} + 8\ CH_3OH \longrightarrow 4\ (CH_3O)_2\overset{\displaystyle S}{\overset{\|}{P}}SH + 2\ H_2S$$

phosphorus   methyl   dimethyl   hydrogen
pentasulfide   alcohol   phosphoro-   sulfide
                                     dithioic acid

$$(CH_3O)_2\overset{\displaystyle S}{\overset{\|}{P}}\!\!-\!\!\underset{H}{S} + \underset{\underset{\text{diethyl maleate}}{CHCOOC_2H_5}}{\overset{CHCOOC_2H_5}{|}} \longrightarrow (CH_3O)_2\ \overset{\displaystyle S}{\overset{\|}{P}}\!\!-\!\!S\!\!-\!\!\underset{CH_2COOC_2H_5}{\overset{CHCOOC_2H_5}{|}}$$

malathion

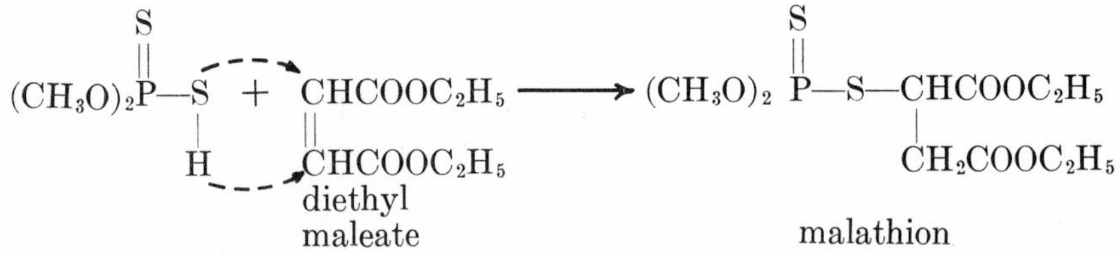

For Thimet, the diethyl phosphorodithioic acid is prepared by the reaction of phosphorus pentasulfide with ethyl alcohol. Diethyl phosphorodithioic acid then reacts with formaldehyde and ethyl mercaptan to eliminate one mole of water and give Thimet:

Thimet:

$$P_4S_{10} + 8\ C_2H_5OH \longrightarrow 4\ (C_2H_5O)_2\overset{\displaystyle S}{\overset{\|}{P}}SH + 2\ H_2S$$

phosphorus   ethyl   diethyl   hydrogen
pentasulfide   alcohol   phosphoro-   sulfide
                                     dithioic acid

$$(C_2H_5O)_2\overset{\displaystyle S}{\overset{\|}{P}}SH + \overset{\displaystyle O}{\overset{\|}{HCH}} + HSC_2H_5 \longrightarrow (C_2H_5O)_2\overset{\displaystyle S}{\overset{\|}{P}}SCH_2SC_2H_5 + H_2O$$

            formaldehyde   ethyl   Thimet
                        mercaptan

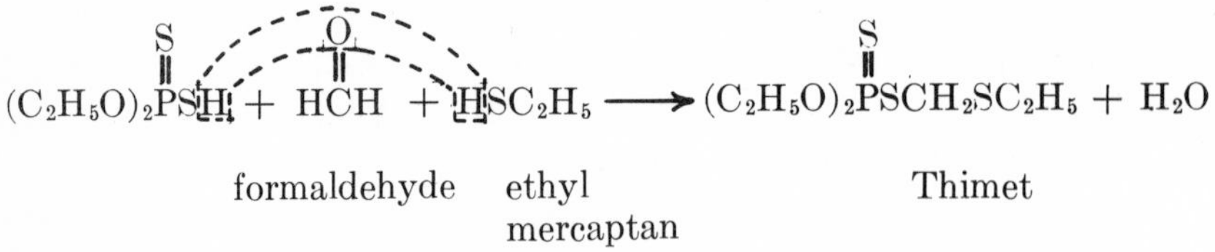

Trithion, as shown on p. 165, is also prepared from diethyl phosphorodithioic acid $[(C_2H_5O)_2\overset{\displaystyle S}{\overset{\|}{P}}SH]$. Many other important insecticides are prepared from the diethyl or dimethyl phosphorodithioic acid which in turn are made by using phosphorus pentasulfide as the intermediate. The above three examples are illustrations.

Trithion:

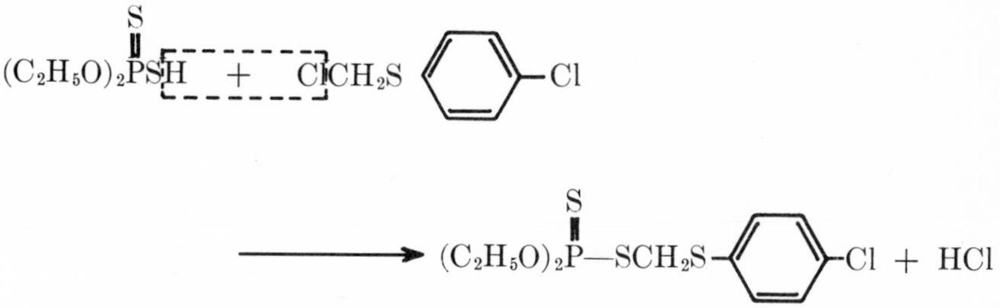

$$(C_2H_5O)_2\overset{\displaystyle S}{\overset{\|}{P}}SH \; + \; ClCH_2S{-}\!\!\bigcirc\!\!{-}Cl$$

$$\longrightarrow (C_2H_5O)_2\overset{\displaystyle S}{\overset{\|}{P}}{-}SCH_2S{-}\!\!\bigcirc\!\!{-}Cl \; + \; HCl$$

Dialkyl phosphorodithioic acids can be chlorinated to form dialkyl phosphorochloridothionate $[(RO)_2\overset{\displaystyle S}{\overset{\|}{P}}Cl]$:

$$(RO)_2\overset{\displaystyle S}{\overset{\|}{P}}SH \qquad + \; Cl_2 \; \longrightarrow \; (RO)_2\overset{\displaystyle S}{\overset{\|}{P}}Cl + HCl + S$$
dialkyl   chlorine   dialkyl
phosphorodithioic acid   phosphoro-
chlorido-
thionate

These dialkyl phosphorochloridothionates are also intermediates in the synthesis of some well known insecticides. For example, the dimethyl phosphorochloridothionate $[(CH_3O)_2\overset{\displaystyle S}{\overset{\|}{P}}Cl]$, is the intermediate for methyl parathion:

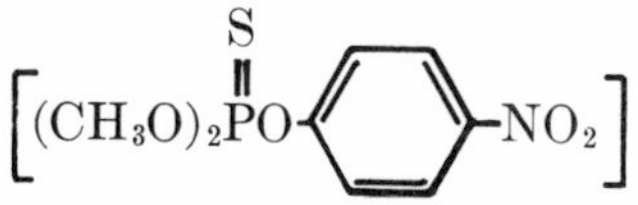

$$\left[(CH_3O)_2\overset{\displaystyle S}{\overset{\|}{P}}O{-}\!\!\bigcirc\!\!{-}NO_2 \right]$$

It forms methyl parathion by the reaction with sodium *p*-nitrophenolate,

$$NaO{-}\!\!\bigcirc\!\!{-}NO_2,$$

as shown in the following equation:

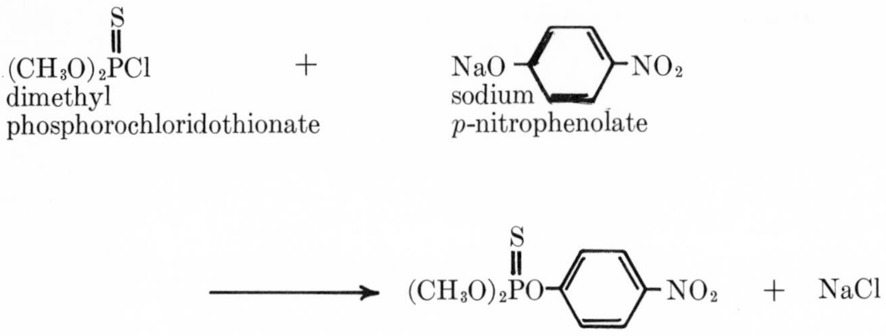

$$(CH_3O)_2\overset{\overset{\displaystyle S}{\|}}{P}Cl \quad + \quad NaO\!-\!\!\bigcirc\!\!-\!NO_2$$

dimethyl   sodium
phosphorochloridothionate   p-nitrophenolate

$$\longrightarrow \quad (CH_3O)_2\overset{\overset{\displaystyle S}{\|}}{P}O\!-\!\!\bigcirc\!\!-\!NO_2 \quad + \quad NaCl$$

methyl parathion

Similarly, diethyl phosphorochloridothionate $[(C_2H_5O)_2\overset{\overset{\displaystyle S}{\|}}{P}Cl]$ is used as the intermediate to make parathion:

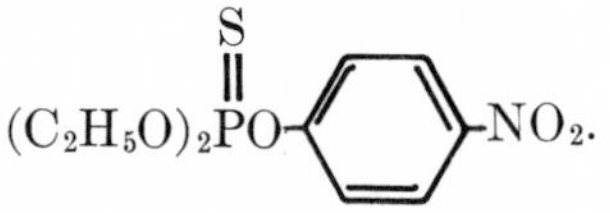

$$(C_2H_5O)_2\overset{\overset{\displaystyle S}{\|}}{P}O\!-\!\!\bigcirc\!\!-\!NO_2.$$

The parathions are used extensively to control cotton pests.

LUBRICANT ADDITIVES. Although other salts have been investigated, it is the zinc salt of dialkyl phosphorodithioic acid which is particularly effective as a lubricant additive. The sodium salts are ineffective; the stannous (tin) salt is very good, but it is too expensive for this use (tin is much more expensive than zinc). The zinc salts are prepared by the reaction of zinc oxide (ZnO) with the dialkyl phosphorodithioic acid $[(RO)_2\overset{\overset{\displaystyle S}{\|}}{P}SH]$:

$$ZnO \quad + \quad 2\ (RO)_2\overset{\overset{\displaystyle S}{\|}}{P}SH \quad \longrightarrow Zn[S\overset{\overset{\displaystyle S}{\|}}{P}(OR)_2]_2 \quad + \quad H_2O$$

zinc   dialkyl   zinc dialkyl
oxide   phosphorodithioic acid   phosphorodithioate

The nature of the alkyl groups depends largely on the final application of the compound. Isoamyl, hexyl, or cyclohexyl derivatives have all been used for more than 20 years.

The use of zinc dialkyl phosphorodithioate as an anti-wear additive in lubricating oil is related to the process of coating the surface of the metal by phosphatization, as discussed in Chapter 9. In a new engine, the metal surface is coated with a phosphatic surface that reduces wear on the metal surfaces in the engine parts during break-in. When the engine is running, this phosphatic coating is gradually replaced by another coating from the reaction of zinc dialkyl phosphorodithioate with the metal surface; this new coating does not chip off. In bearings, gears, and valve trains, a smooth surface greatly reduces frictional wear and prolongs their useful life.

The function of the zinc dialkyl phosphorodithioate as an extreme pressure additive in gear lubricants is related to its anti-wear properties. In an engine's gear system, the pressure at the point of contact between the two gears is so high that it can actually extrude all of the lubricant from this point. This results in direct metal-to-metal contact, resulting in high wear. The phosphate additive coats the surface of the metal by chemical reaction; the organic portion of the phosphate additive holds the oily lubricant through chemisorption.

The protective coating formed by zinc dialkyl phosphorodithioate also retards corrosion of the bearing surface by the oxidation products of the oil. Further, the dithioate additives prevent or reduce (through chemical reactions) the formation of the corrosive decomposition products in the lubricating oil. This latter reaction is called the "anti-oxidant" reaction and helps to maintain the quality of oil during long usage. Through the use of additives such as these, crankcase oil changes are reduced. The formulating of lubricants for use in crankcases and gear boxes is a highly developed science. The use of dialkyl phosphorodithioates, prepared from phosphorus pentasulfide represents just a part of this development. Non-phosphorus derivatives are also important in a lubricant formulation.

ORE FLOTATION AGENTS. Flotation agents are compounds used to concentrate desired minerals in crude ores. In mining copper, for example, the copper sulfide ore is contaminated with large amounts of sand, clay, and other minerals. Separation of the ore from these contaminants efficiently and economically is necessary.

The ore flotation process is based on the property of a properly treated surface to have a high affinity for air bubbles and a low affinity for water. In copper sulfide flotation, for example, finely ground ore is treated in a water slurry with such compounds as the sodium or ammonium salts of dialkyl phosphorodithioic acid (called collectors). These salts selectively wet the surface of the copper sulfide particles but do not wet the surface of the contaminants. Air is then bubbled into the water slurry. The air bubbles attach themselves to the treated surface of the copper sulfide and not to the contaminants. The air bubbles with the copper sulfide attached rise as a froth to the top and are then separated from the contaminants which have been wetted by the water and have settled to the bottom. The concentration of dialkyl phosphorodithioic acid collector used for flotation is only about 0.01 to 0.1 pound per ton of ore.

Flotation is used to recover such other valuable minerals as molybdenum, lead, zinc, silver and gold, as well as non-metallic minerals like mica and feldspar. The particular flotation collector used depends on the mineral to be collected. Generally, most of the dialkyl phosphorodithioates used for this purpose are prepared from ethyl, isopropyl and sec-butyl alcohols. The important factor is high selectivity for the preferential coating of the specific mineral for air flotation. Other agents such as the dithiocarbonates and xanthates are also used as collectors. However, phosphorodithioates prepared from the reaction of phosphorus pentasulfide and alcohols are also very important.

15

The Code of Life—DNA and RNA

One of our more enduring nursery rhymes proposes that little girls are made of sugar and spice and everything nice and that little boys are made of snakes and snails and puppy dogs' tails. The unromantic chemist, however, sees us as various molecular arrangements of the following elements:

	%
Oxygen	65
Carbon	18
Hydrogen	10
Nitrogen	3
Calcium	1.5
Phosphorus	1
Other	1.5

To shatter our illusions further, it has been calculated that the net value of the chemical ingredients in the human body is about $5.60. This figure is for a 150-pound person and is based on 1974 prices. In 1936, during the depression, the human body was worth only $.98.

Oxygen, carbon, and hydrogen constitute the greatest percentage of the body. Most of the hydrogen and oxygen, of course, is present as water. Phosphorus represents only 1% of our body weight, but, as we shall see, a very important 1%.

About 23% of the human skeleton is mineral matter. The phosphorus content of this mineral matter, calculated as tricalcium phosphate, $Ca_3(PO_4)_2$, represents 87% of the total. Similarly, our teeth are basically calcium phosphates.

The proper biological functioning of our bodies depends on the action of the many phosphorus-containing compounds, or biophosphates, in our systems. The most important are DNA (deoxyribonucleic acid) and RNA (ribonucleic acid). These compounds carry the genetic code which determines what we are. Another important biochemical is ATP, or adenosine triphosphate, the energy carrier. The chemistry and some of the functions of DNA and RNA are discussed in this chapter. Phosphorus energy transfer agents such as ATP are covered in the next chapter.

Phosphorus enters our bodies mainly through our food. Every bite of meat or mouthful of vegetables contains some phosphorus-containing compounds since phosphorus is also essential to plants and animals. Plants take up phosphorus from soil nutrients (fertilizers, decayed organic matter, etc.); animals eat plants and we eat animals and plants. We also consume phosphorus compounds such as calcium phosphates (discussed in earlier chapters), which are incorporated as supplemental nutritive additives in food. Some is excreted and ends up in streams, rivers, and lakes, where plant life such as algae utilize it as one of their 16-17 required nutrients. The excess growth of algae leads to eutrophication, a subject which was covered in Chapter 8.

"Knock it off, Jenkins."

What role does phosphorus-containing DNA, deoxyribonucleic acid, play in our biological system? When an egg cell is fertilized by a sperm cell, a new life begins. The fertilized egg absorbs various nutrients from its surroundings and begins to grow in an orderly manner. Different eggs grow into different animals—e.g., some eggs hatch into chickens, others grow into elephants or monkeys, and some become human beings. Even humans show differences. Some are male; some are female. Some become pretty blondes; some brunettes. The nutrient molecules or raw materials available to fertilized eggs are chemically not very different. What then determines how these raw materials are arranged into a unique final shape?

The answer came after years of accumulated research by many scientists. According to present theory, small units in the fertilized egg called genes transmit hereditary characteristics. The active components in the gene which serve as templates for growth and differentiation are the DNA molecules.

DNA is a very large molecule. Its presently accepted basic structure was established in 1953 by James B. Watson, Maurice H. F. Wilkins, and Francis H. C. Crick. (Their work so advanced the science of molecular biology that they were awarded a Nobel prize in 1962. The exciting story of the unravelling of the true DNA structure is told in "The Double Helix," by James Watson.) As proposed, a DNA molecule consists of two very long thin chains, twisted around each other as a regular double helix. This spiral-staircase arrangement is shown in Figure 1. These chains, called nucleic acids, are formed by the combination of smaller building blocks called nucleotides. A nucleotide in turn is a combination of a nitrogen-containing base, a five-carbon sugar called deoxyribose, and

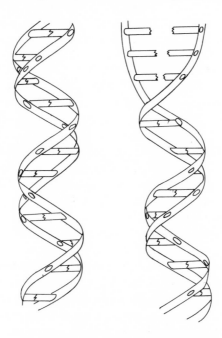

(a)   Coiled   (b) Partially uncoiled

*Figure 1.   Pictorial representation
of DNA double helix*

phosphoric acid. There are four different nitrogen-containing bases in a
DNA molecule: adenine, cytosine, guanine, and thymine. A nucleotide
with adenine as the nitrogen-containing base is shown in Figure 2.

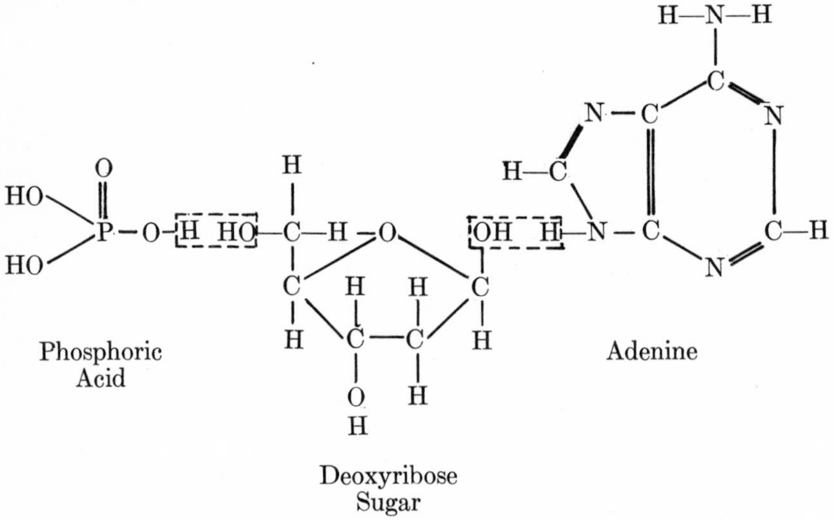

Phosphoric
Acid

Deoxyribose
Sugar

Adenine

Figure 2.   A nucleotide, from its three components

That portion of the nucleotide which contains only the nitrogen base
and the deoxyribose sugar is called the nucleoside. From the standpoint
of phosphorus chemistry, we can visualize phosphoric acid as having the
important function of joining the thousands of nucleosides together to
form the long chain of nucleic acid. A small section of a nucleic acid
chain is shown in Figure 3. Although the four different nitrogen-containing bases are shown, in an actual nucleic acid chain, the arrangement of
these bases is specific for a specific nucleic acid. It is the particular
arrangement of these bases at various positions in the nucleic acid chain
that is the basic genetic code of life.

In Figure 1, both sides, or banisters, of the staircase are identical
since each is a long chain of alternate phosphate and deoxyribose molecules. Each step consists of two nitrogen bases linked together. One base
is connected to deoxyribose on the left banister, while the other is connected to the deoxyribose on the right banister. Of the four nitrogen
bases in a nucleic acid chain, because of their physical size, the base
thymine in one chain always bonds with the base adenine in the other
chain, and the base cytosine always bonds with the base guanine. In
other words, thymine is complementary to adenine, and cytosine is complementary to guanine. Because the nucleic acid chain is so long, many

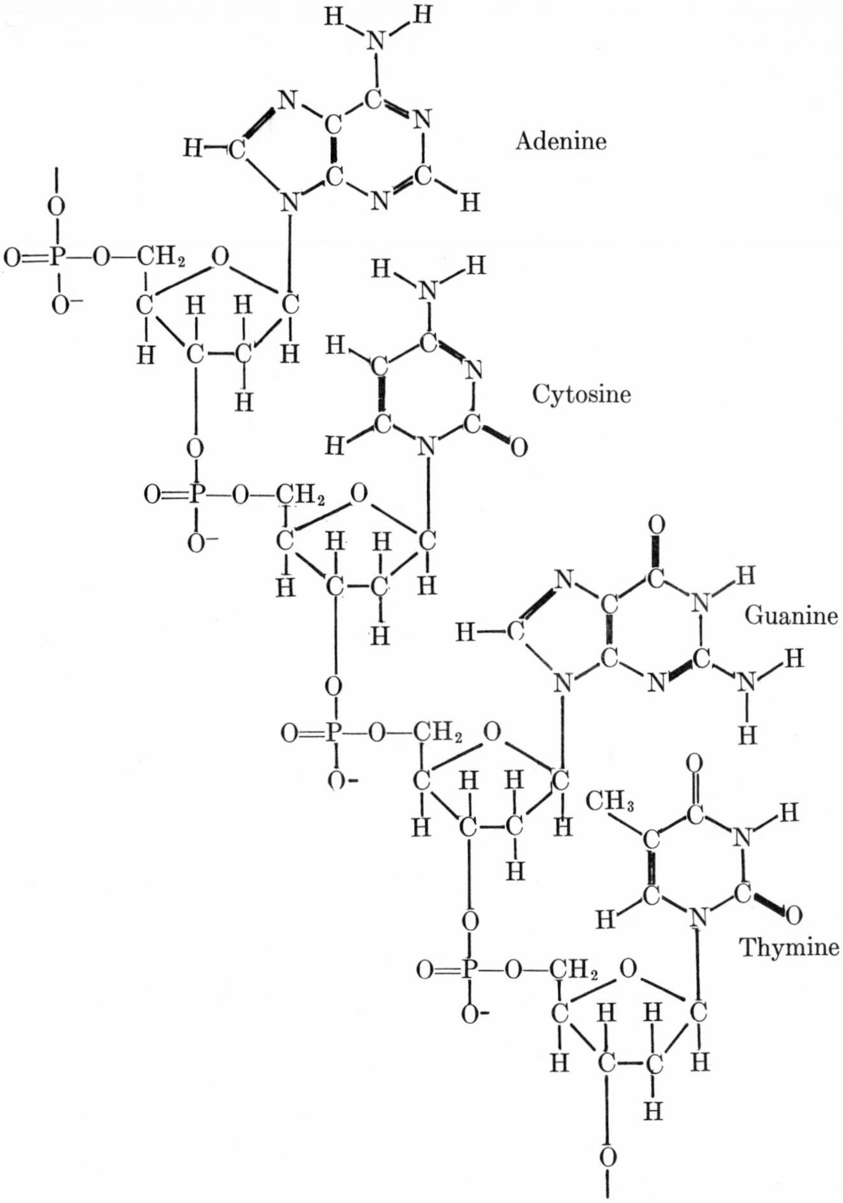

Figure 3. A small section of the DNA chain

different combinations and arrangements of these bases are possible. The difference between man and fish, frog and yeast, male and female, depends on the percentage of each base and its specific position in the

DNA molecule. Heredity information is transmitted *via* these arrangements of bases in DNA.

The present view is that the bases are bonded together by fairly weak hydrogen bonds, as shown in Figure 4. When reproduction begins,

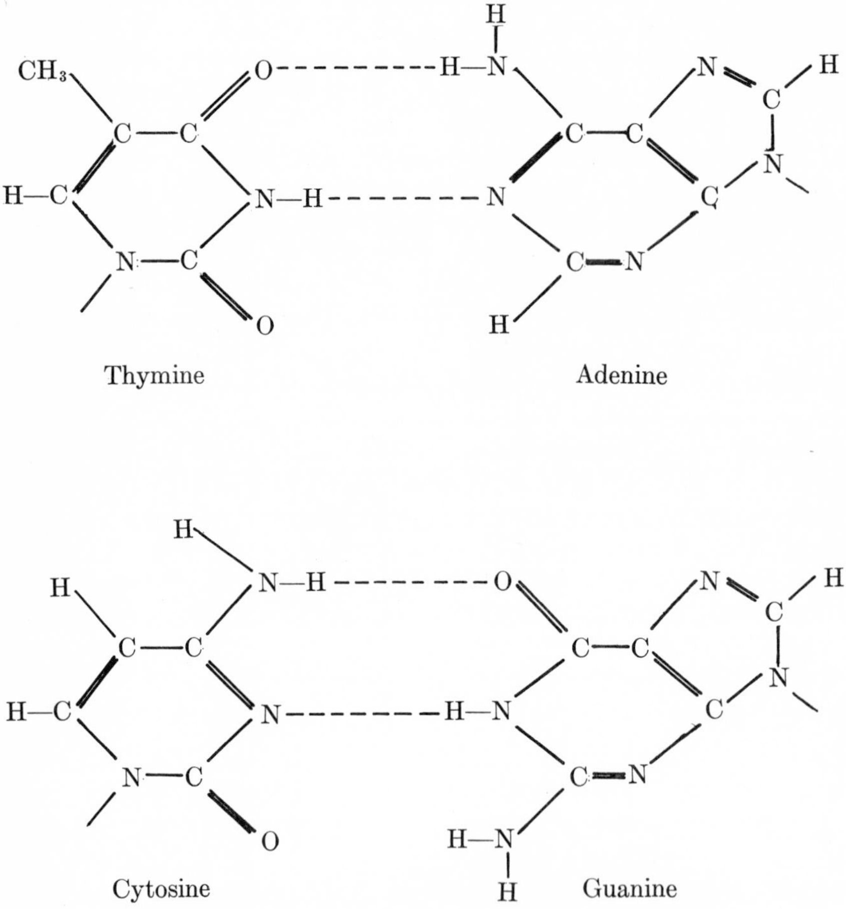

Figure 4. Hydrogen bonds (dotted lines) between complementary bases linking two helical DNA chains together

these hydrogen bonds rupture, and the two DNA chains separate (Figure 1b). Each chain uncoils to form two independent single helices. Each helix is now a template for arranging and organizing the smaller nucleotide units present in the cell into a complementary chain. The left chain

is the template according to which the cell synthesizes a new complementary right chain, while the right chain does the same for the new left
chain. Thus a new identical double helix of nucleic acid chains is synthesized from the original.

Since a DNA chain contains thousands of nucleotide units and reproduction goes on in billions of living cells each second, does nature ever
make a mistake, so that the daughter DNA double helix produced is not
exactly like the mother DNA? According to Jacques Monod, Nobel prize-
winner for medicine and physiology in 1965, some errors do occur, and
when they do, the new DNA molecule with the slight difference will
produce a cell with the new characteristic. Since DNA molecules generally produce daughter molecules identical to themselves, once an error
occurs, subsequent DNA generations and their respective cells will contain the differences caused by the original error. This invariance to
change is retained until the next error ocurrs. It is believed that mutations
occur as the result of such errors, and that it took a succession of many
errors and coincidences for the living cell to evolve into man as we know
him today. This theory is compatible with Darwin's theory of evolution.
According to him, the evolution of a living creature is determined by
natural selection—*i.e.*, survival of the fittest. Many past errors of DNA
reproduction fizzled out because the mutation produced couldn't survive
its hostile environment.

In addition to DNA, the other important phosphorus-containing
molecule which is essential to the biological process is RNA, ribonucleic
acid. RNA's structure is similar to DNA's, with the backbone of the helix
(banister of the spiral staircase) constructed of ribose sugar molecules
(instead of deoxyribose sugar) joined by phosphoric acid. Ribose differs
from deoxyribose in having only one oxygen atom at carbon 2, as shown
below:

ribose   deoxyribose

Also, one of its four nitrogen bases is uracil rather than the thymine base
in DNA.

RNA exists mostly as a single polynucleotide strand or single helix. However, when a small section exists in a double helix structure, the steps in the very short spiral staircase section are also formed by complementary base pairs. As in DNA, cytosine is bonded to guanine, but adenine, instead of bonding to thymine, is bonded to uracil, as shown below:

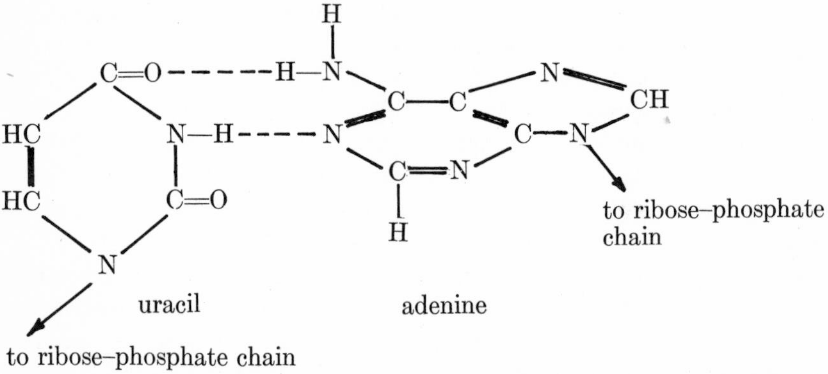
uracil
adenine
to ribose–phosphate chain
to ribose–phosphate chain

Many of the functions of RNA in a biological system are still undiscovered. However, one of its major functions is as the template for synthesizing protein molecules from amino acids present in the body. An amino acid is an organic compound containing an amino group, $-NH_2$, and a carboxylic acid group, $-COOH$. For example, the amino acid glycine has the structure $H_2N\text{–}CH_2\text{–}COOH$, and the amino acid alanine has the structure $H_2N\text{–}\overset{\overset{\displaystyle CH_3}{|}}{C}HCOOH$. When two amino acids join and eliminate a water molecule, an amide linkage forms:

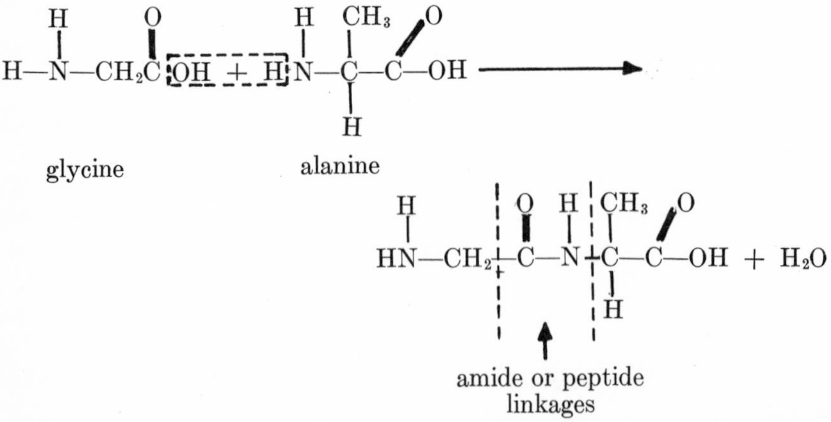
glycine
alanine
amide or peptide linkages

This amide linkage is also called a peptide bond. About 20 different amino acids are commonly used to build a protein molecule. When hundreds of these amino acids are joined, they form a polypeptide or a protein molecule. In a protein molecule the amino acids are connected in a preordered sequence which depends on the final function of the protein. For example, liver cells make protein suitable for use in the liver and nowhere else. The liver protein has chain length, amino acid sequence, and geometric structure different from those proteins made for the pancreas. Even enzymes, the catalysts which accelerate biochemical reactions, are protein molecules. They differ from each other with respect to size, amino acid sequence, and three-dimensional geometric structure.

How do cells make protein molecules with these differences? The gross mechanism for protein synthesis has been elucidated by molecular biologists through some very ingenious laboratory experiments, but some details of this mechanism are still not clearly understood. Apparently, however, protein synthesis involves three distinct types of RNA molecules:

(1) Ribosomal RNA, or rRNA. When combined with a protein, it is called a ribosome. Ribosomes are located in the cytoplasm of the cell surrounding the nucleus, and they are the factories where proteins are made.

(2) Messenger RNA, or mRNA. As the name indicates, mRNA (a phosphorus-containing nucleic acid) carries the code from the DNA which determines how the protein should be made. The amino acid sequence of the protein is dictated by this code.

(3) Transfer RNA, or tRNA. Molecules of tRNA are relatively small, containing only about 80 phosphorus-containing nucleotides. Each tRNA carries one specific amino acid and three specific consecutive nucleotides with their three bases; these bases serve as a code word which fits exactly with three consecutive complementary nitrogen bases along the long mRNA nucleotide strand.

When the body needs a certain protein, a signal is sent to the cell. Inside the cell nucleus, specific enzymes come into play, and the DNA double helix, with its coded messages, uncoils. Portions of the two strands separate. The single nucleotides present in the cytoplasm of the cell attach themselves quickly to one of the DNA strands *via* complementary nitrogen bases and form a single strand of mRNA. This mRNA is thus a long chain of nucleotides with nitrogen bases arranged in a specific sequence according to the complementary base sequence in the original DNA strand. The message carried by the sequential arrangement of nitrogen bases in the original DNA is passed on to the new mRNA strand. Every three consecutive bases represent a code word for a specific amino acid. After the mRNA molecule is formed, it leaves the nucleus and attaches itself to a ribosome. Meanwhile the separated portion of DNA double helix in the nucleus rewinds and is available for later use. After it is picked up by the ribosome, the mRNA strand begins to call for the proper amino acids according to the coded message it contains.

The tRNA, which carries the called-for amino acid and its three nucleotides with the three-nitrogen base code word for that amino acid, answers the call. It moves in and lodges on the three consecutive nucleotides with the complementary bases on the mRNA strand. The mRNA then calls for the next tRNA with its amino acid. It comes into the site adjacent to the first amino acid, and the two amino acids join to form

Reproduced by permission of Janet Carver

"So what if your chemistry teacher told you you're only oxygen, carbon, hydrogen, nitrogen, calcium, and phosphorus—you're worth more than $4.00 [$5.60] to me!"

a peptide bond. The tRNA for the first amino acid leaves, and the second tRNA now has two amino acids attached to it. The next three-base code word on the mRNA strand calls for another tRNA with its specific amino acid, and the three complementary bases to move into the ribosome to be joined with the first two amino acids carried by tRNA already there. The new tRNA displaces the old tRNA. It now carries three bonded amino acids and waits for the fourth tRNA to bring in its specific amino acid. By repeating this process, amino acids are brought to the ribosome in the sequence dictated by the coded message from the mRNA, and they rapidly build into a long protein chain. When completed, the protein chain is released into the cytoplasm, where it rearranges into its proper geometric structure for use. The freed mRNA when attached to another ribosome can be used again to assemble another protein molecule.

Although this description of protein synthesis is simplified, it shows how important DNA and RNA are in the biological system. Once the nature and function of these phosphorus compounds are understood, scientists should be able to unlock the secret of life itself.

More detailed and sophisticated discussion of the molecular biology of DNA and RNA is available in the original publications of such distinguished scientists as E. F. Hoppe-Seyler and F. Miescher, who first isolated nucleic acid in 1869; Oswald Avery and his co-workers, who in 1943 resolved the role of DNA as a carrier of genetic information; J. B. Watson, M. H. F. Wilkins, and F. H. C. Crick, who established the presently accepted double helix structure for DNA; F. H. C. Crick and his co-workers, who in 1961 provided experimental evidence to support the concept of triplet code; S. Ochoa, who in 1955 discovered a catalyst to manufacture synthetic RNA; and M. W. Nirenberg and J. H. Matthaei, who in 1961 used Ochoa's synthetic RNA process to crack the genetic code for making polypeptide. The publications of these scientists will lead you to additional important contributions by other molecular biologists.

A good summary of the functions of DNA and RNA and some of the research done by various scientists to arrive at the understanding of these functions are described in the books cited below (*1, 2*).

Literature Cited

1. Louis Levine, "Biology of the Gene," The C. V. Mosby Company, St. Louis, 1969.
2. Adrian M. Srb, Ray D. Owen, Robert S. Edgar, "General Genetics," 2nd ed., W. H. Freeman and Co., San Francisco, 1965.

16

ATP—The Energy Carrier

The synthesis of DNA, RNA, and protein molecules requires work or energy. For that matter, most human activity—breathing, eating, running, playing tennis or golf—requires energy. Energy is also required when a plant synthesizes sugar or starch molecules or when a firefly emits light. The source of energy for all these activities is the biotriphosphates. Energy is given off when one triphosphate bond is broken to form a monophosphate and a diphosphate.

For most biological processes, the more important energy carrier or energy transfer agent in the cell is a biotriphosphate called adenosine triphosphate, or ATP (Figure 1).

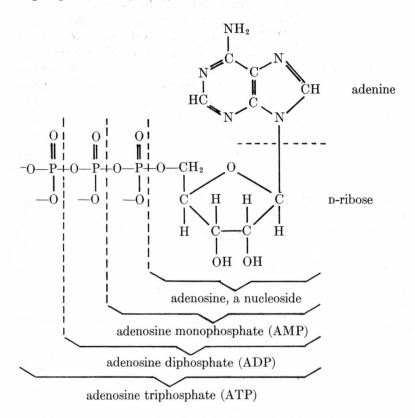

Figure 1.   *Adenosine triphosphate (ATP, the energy carrier)*

As shown in Figure 1, ATP is composed of three kinds of building blocks: the base adenine, the five-carbon sugar D-ribose (these two, when combined, form one of the nucleosides for RNA), and a chain of three phosphate groups. The phosphate groups are connected by P–O–P bonds known as anhydride linkages. (ADP, or adenosine diphosphate, also an important biological phosphorus compound, has only two phosphate groups.) ATP resides in cells in concentrations between 0.5 and 2.5 mg/cc. When one phosphate group in ATP is hydrolyzed off at neutral pH at 25°C to give ADP and an orthophosphate, 7000 calories of energy are liberated. This free energy of hydrolysis is used to perform biological work.

Table I.   Free Energy of Hydrolysis of Biophosphate Compounds

	cal/mole
phosphoenolpyruvate	−12,800
1,3-diphosphoglycerate	−11,800
phosphocreatine	−10,500
acetyl phosphate	−10,100
ATP	−7,000
glucose 1-phosphate	−5,000
fructose 6-phosphate	−3,800
glucose 6-phosphate	−3,300
3-phosphoglycerate	−3,100
glycerol 1-phosphate	−2,300

ATP is not the only phosphorus energy carrier in the biological system. Others are shown in Table I. Some have a higher free energy of hydrolysis than ATP, some lower. The minus sign indicates that the system loses or gives off energy on hydrolysis.

A phosphate group from a compound with a higher free energy of hydrolysis can be transferred to form phosphate compounds with lower free energies of hydrolysis. With each phosphate transfer, energy is also transferred.

How do organic phosphates with free energy of hydrolysis obtain their energy? In particular, how does ATP obtain its bond energy? The answer lies in the food we eat. In general, the breakdown of food molecules in the cell is always coupled with the eventual formation of ATP from ADP and monophosphate or from orthophosphate. Two examples show that energy released by this breakdown is partially conserved by the formation of ATP from ADP and an orthophosphate. Both examples involve the breakdown of a glucose (six-carbon sugar) molecule. One is by an anaerobic (absence of oxygen) cell, the other by an aerobic cell.

Anaerobic Oxidation of Glucose

One molecule of glucose in a living cell breaks down under anaerobic conditions to form two molecules of lactate, a three-carbon compound, by the process called glycolysis. This mechanism also involves the net formation of two moles of ATP from two moles of ADP and two moles of orthophosphate, as shown in the following overall reaction:

$$\text{glucose} + 2HPO_4^{2-} + 2ADP^{3-} \longrightarrow 2 \text{ lactate}^- + 2ATP^{4-} + 2H_2O$$

Simple breakdown of glucose to lactate is accompanied by a liberation or change in free energy of $-52,000$ calories per mole. Since two moles of ATP are formed from ADP and orthophosphate, and each mole of ATP requires at least 7000 calories, the summation is $14,000/52,000 \times 100 = 27\%$. In other words, about 27% of the energy released when a molecule of glucose breaks down to lactate is conserved as ATP. The lactate then leaves the cell as waste.

The anaerobic breakdown of glucose to two lactates is known as the Embden–Meyerhof cycle, named after two German biochemists who predicted and experimentally proved the overall pattern in the 1930's. Its mechanism is not as simple as shown. It requires 11 recognizable steps and involves 11 specific enzyme catalysts. Also, each step involves a phosphorylation reaction—that is, all of the intermediates in the glycolysis reaction sequence require the participation of a phosphate group to form an ester of phosphoric acid. The 11 steps are shown in the following simplified equations. The enzyme which catalyzes each reaction is shown above the arrow.

These sequences show that even the anaerobic breakdown of glucose requires energy, and this energy is supplied by ATP. Equations 1 and 3 each show that energy from one mole of ATP is needed. Thus energy

1. 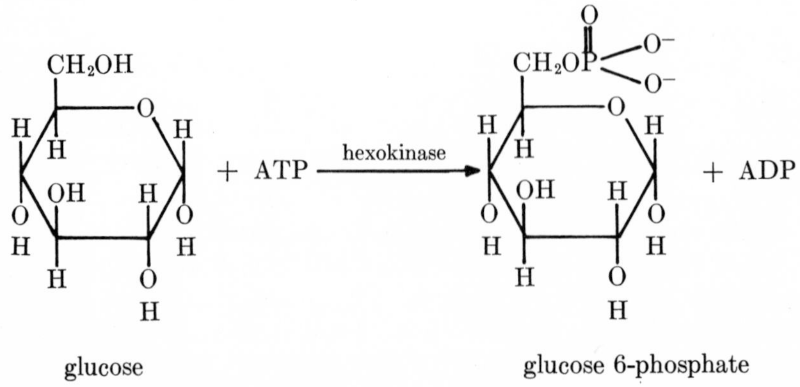   glucose   glucose 6-phosphate

2.

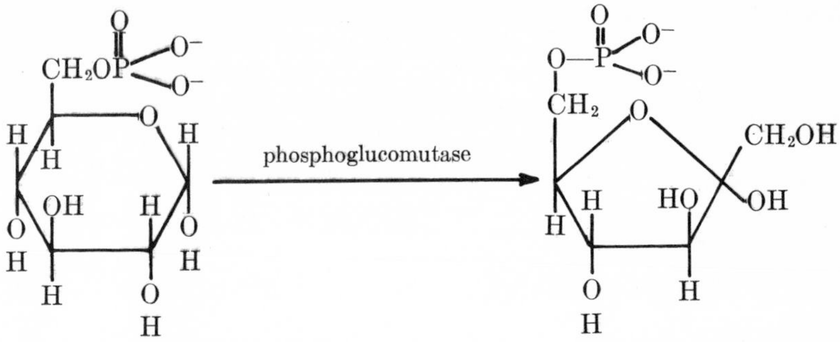

glucose 6-phosphate   fructose 6-phosphate

3.

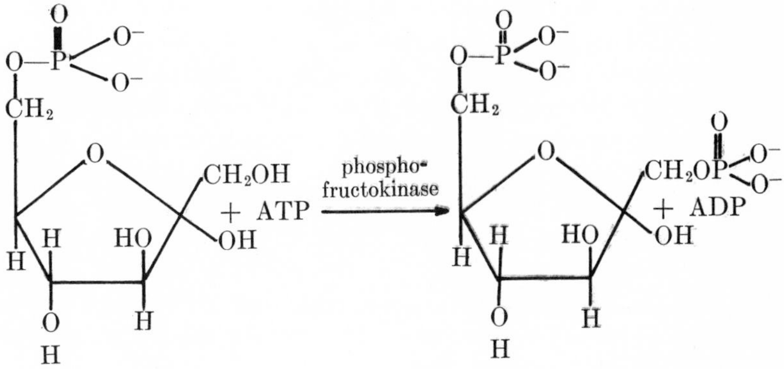

fructose 6-phosphate   fructose 1,6-diphosphate

4.

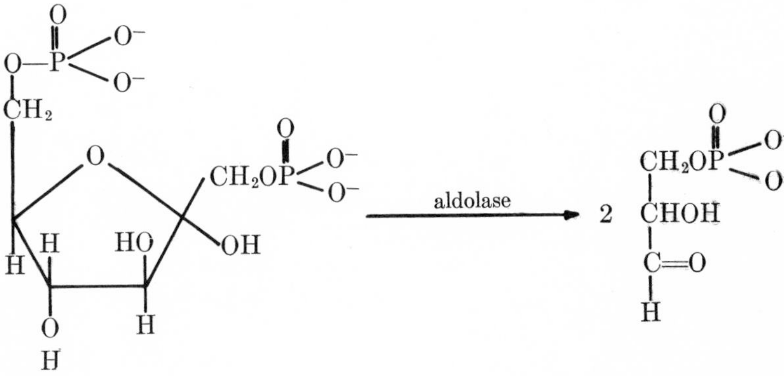

fructose 1,6-diphosphate   3-phosphoglyceraldehyde

5. 2 3-phosphoglyceraldehyde $\xrightarrow{\text{triose phosphate isomerase}}$ 2 dihydroxyacetone phosphate

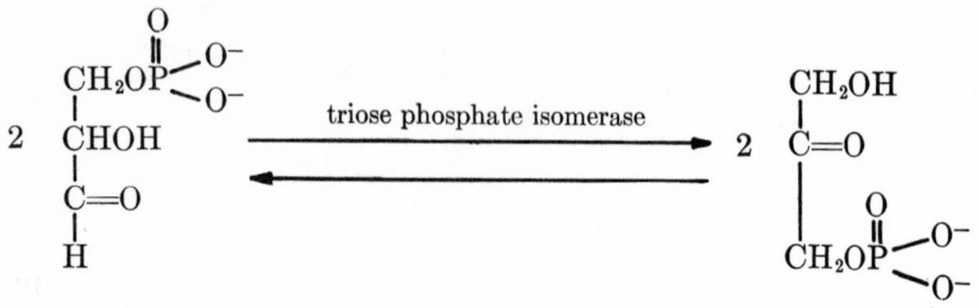

3-phosphoglyceraldehyde   dihydroxyacetone phosphate

6. 2 3-phosphoglyceraldehyde $+ 2$ ... $+ 2$ NAD$_{ox}$

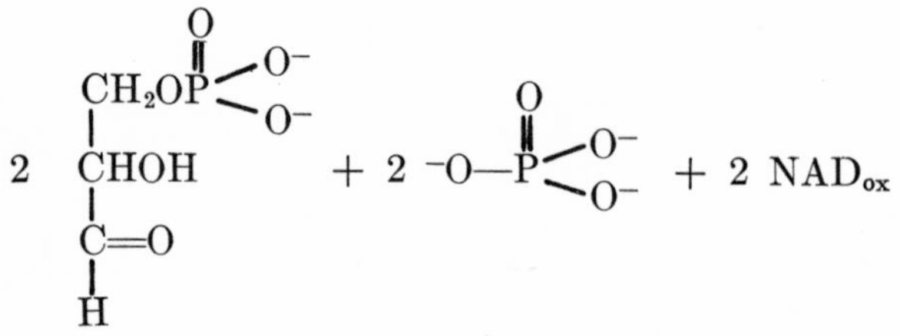

3-phosphoglyceraldehyde

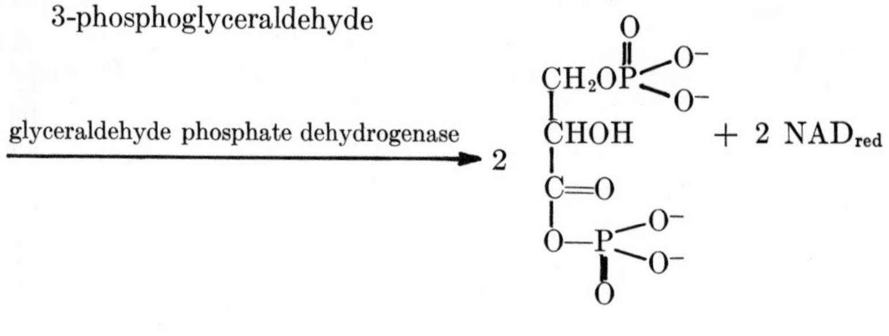

glyceraldehyde phosphate dehydrogenase $\longrightarrow$ 2 ... $+ 2$ NAD$_{red}$

1,3-diphosphoglycerate

7. 2 1,3-diphosphoglycerate $+ 2$ ADP $\xrightarrow[\text{kinase}]{\text{diphosphoglycerate}}$ 2 3-phosphoglycerate $+ 2$ ATP

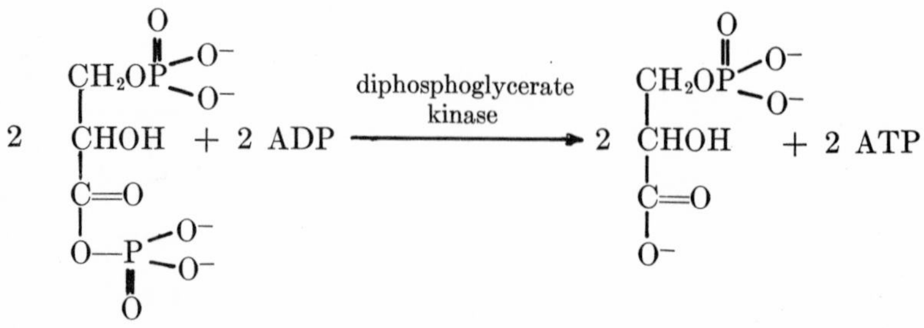

1,3-diphosphoglycerate   3-phosphoglycerate

8. 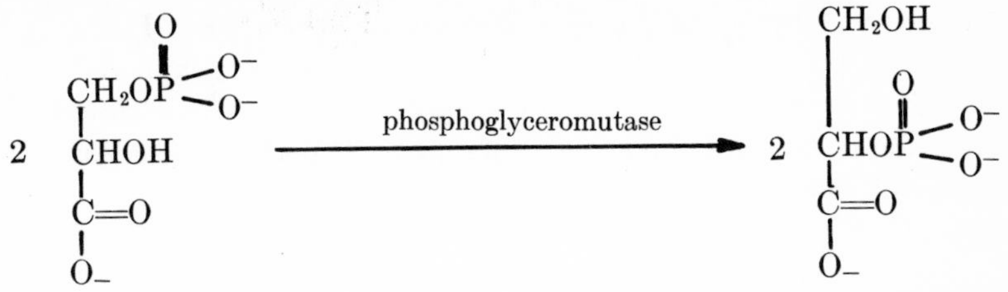   2   3-phosphoglycerate $\xrightarrow{\text{phosphoglyceromutase}}$ 2   2-phosphoglycerate

3-phosphoglycerate   2-phosphoglycerate

9. 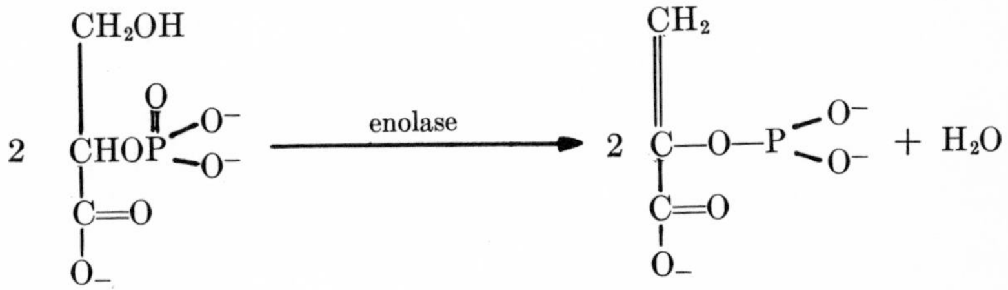   2   2-phosphoglycerate $\xrightarrow{\text{enolase}}$ 2   phosphoenolpyruvate $+ H_2O$

2-phosphoglycerate   phosphoenolpyruvate

10. 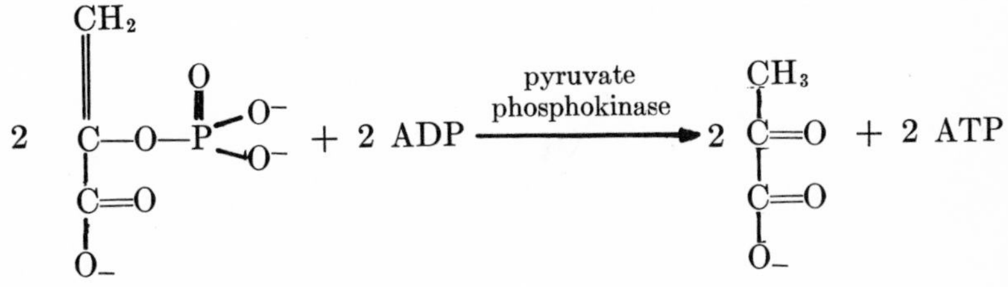   2   phosphoenolpyruvate $+ 2$ ADP $\xrightarrow[\text{phosphokinase}]{\text{pyruvate}}$ 2   pyruvate $+ 2$ ATP

phosphoenolpyruvate   pyruvate

11. 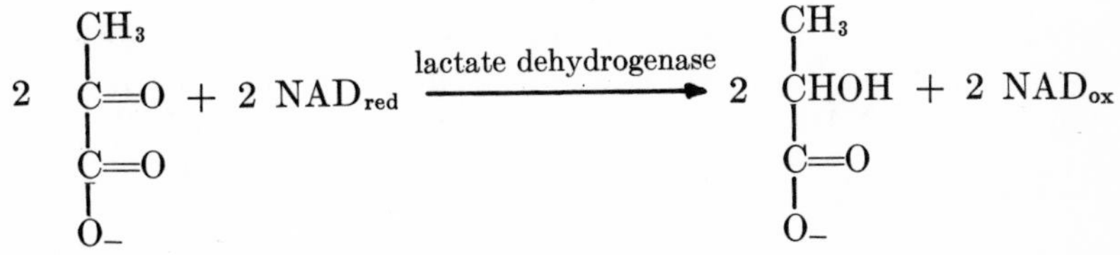   2   pyruvate $+ 2$ NAD_{red} $\xrightarrow{\text{lactate dehydrogenase}}$ 2   lactate $+ 2$ NAD_{ox}

pyruvate   lactate

from two moles of ATP is needed early in the process. However, when the mechanism reaches Equations 7 and 10, four moles of ATP are produced. The net gain in the 11 glycolysis steps is two moles of ATP from the energy supplied when one mole of glucose is converted to two moles of lactate.

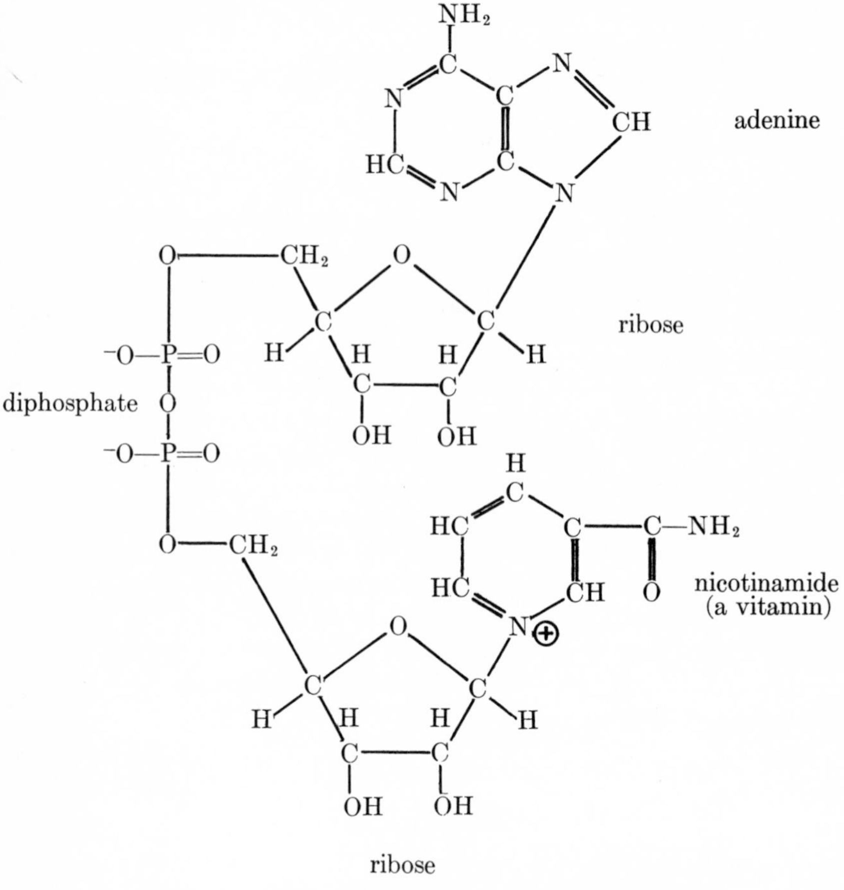

Figure 2.   Nicotinamide adenine dinucleotide (NAD)

Another phosphorus-containing compound which enters this sequence of reactions, *via* Equations 6 and 11, is NAD, nicotinamide adenine dinucleotide (Figure 2). NAD is an electron carrier—*i.e.*, the nicotinamide moiety of the NAD is the electron carrier. In Equation 6, the aldehyde group $-\overset{\overset{\displaystyle O}{\displaystyle \|}}{C}-H$ is oxidized to the carboxylic acid $-\overset{\overset{\displaystyle O}{\displaystyle \|}}{C}-O^-$ with the loss of two electrons per molecule. The oxygen in this oxidation is supplied by

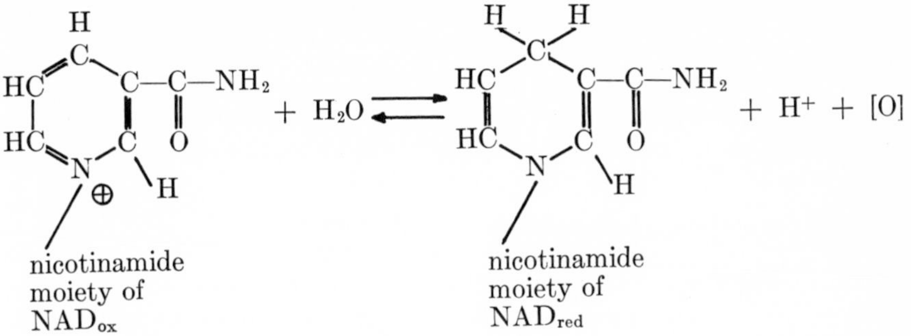

nicotinamide
moiety of
NAD_{ox}

nicotinamide
moiety of
NAD_{red}

a water molecule, and the electrons are removed by the two hydrogen atoms from the same water molecule. For this reason, biochemists also call such an oxidation reaction a dehydrogenation reaction.

Each oxidized molecule of NAD, or NAD_{ox}, can accept two electrons to become reduced NAD, or NAD_{red}. In Equation 11, pyruvate is reduced to lactate, and each molecule reduced requires two electrons. The needed electrons are brought from Equation 6 by NAD_{red}. After NAD_{red} donates the two electrons to Equation 11, it becomes NAD_{ox} and ready to carry electrons again from Equation 6.

Note that the energy resulting from the oxidation of 3-phosphoglyceraldehyde in Equation 6 is first conserved by the formation of 1,3-diphosphoglycerate. As indicated in Table I, the latter compound has a free energy of hydrolysis of $-11,800$ calories per mole. Some of this energy is transferred to the phosphate bond energy in ATP when it donates a phosphate to ADP to form ATP as shown in Equation 7.

One possible reason why 1,3-diphosphoglycerate has such a high free energy of hydrolysis is its structural makeup:

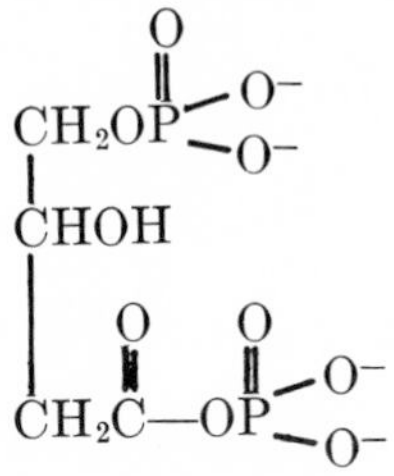

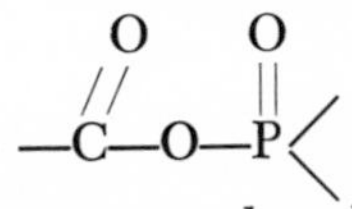

The two phosphate groups close together in the molecule, each with two negative charges, repel each other. Also, the $-C-O-P\langle$ linkage has a high density of electrons. These negative charges repel each other and thus are ready for hydrolysis. After hydrolysis, the negative charges impede the phosphate $\left(\begin{smallmatrix} & O & \\ & \| & O^- \\ -OP & \diagup \\ & \diagdown & O^- \end{smallmatrix}\right)$ and carboxyl groups $\left(\begin{smallmatrix} & O & \\ & \diagup\diagup & \\ C & \diagdown & O^- \end{smallmatrix}\right)$ from recombining. In addition the electrons in these two groups arrange into a more stable configuration, which makes recombination more difficult.

Aerobic Oxidation of Glucose

The anaerobic breakdown of glucose results in the formation of two lactates. Now consider the degradation of glucose by aerobic cells—*i.e.*, cells requiring oxygen. In this case glucose is not converted to lactate but to pyruvate, the product of Equation 10. Pyruvate is then oxidized to CO_2 and H_2O; the total energy released is 686,000 calories. Again, much of this energy is conserved in ATP. Aerobic cell oxidation also depends on enzyme catalysts at each step. These enzymes are fixed in a

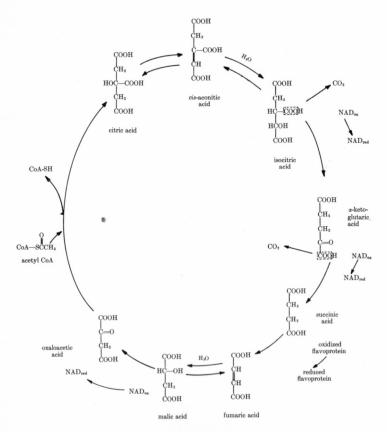

Figure 3.   Krebs tricarboxylic acid cycle

specific geometric arrangement in the mitochondria or the power plants of the cell.

When we eat our daily meals, we are often quite particular about whether the food is properly prepared. We place great emphasis on the flavor and texture. We complain if the meat is tough, the salad not crisp, or the bread stale. As far as the body cells are concerned, however, food is just a source of energy. Body cells oxidize (burn) carbohydrates (from starches and sugars), fatty acids (from fats), and amino acids (from meat and proteins) to yield the energy needed for various functions. The mechanism of this oxidative process is the Krebs tricarboxylic acid cycle (also called the citric acid cycle), is shown in Figure 3. This cycle was deduced and proved in the 1930s by Hans A. Krebs, who was awarded a Nobel prize for this work. According to the Krebs mechanism, food is not converted directly to energy. Preliminary enzymatic reactions break food molecules into two carbon pieces, or acetic acid. The Krebs cycle can accept only acetic acid, $CH_3\overset{\overset{O}{\|}}{C}$—OH, in the form of a derivative of conenzyme A, or CoA, another biophosphate:

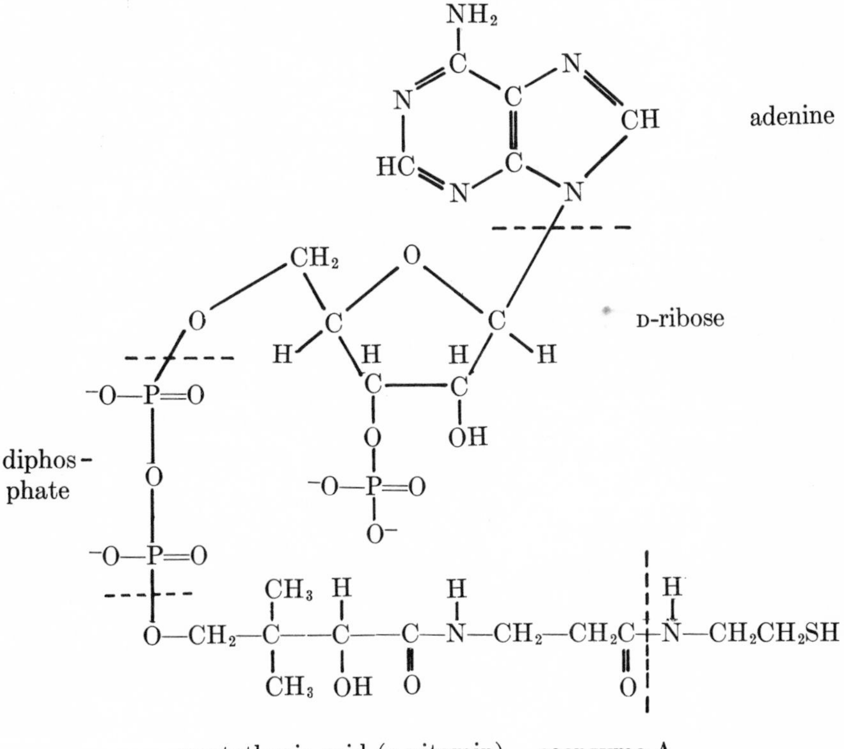

adenine

D-ribose

diphos-phate

pantothenic acid (a vitamin)   coenzyme A

An illustration of coenzyme A as carrier of an acetyl group for the Krebs cycle is shown in the case of acetic acid from the enzymatic oxidation of pyruvic acid of Equation 10 in the glycolysis reactions:

$$CH_3\overset{\overset{\displaystyle O}{\|}}{C}-COOH + H_2O + NAD_{ox} \xrightarrow[\text{dehydrogenase}]{\text{pyruvate}}$$

pyruvic acid

$$CH_3\overset{\overset{\displaystyle O}{\|}}{C}-OH + CO_2 + NAD_{red}$$

acetic acid

Electrons removed from pyruvic acid are accepted by NAD_{ox} and carried in NAD_{red}. The acetic acid formed immediately reacts with CoA to form acetyl CoA:

$$CoA-SH + HOCCH_3 \longrightarrow CoA-S-\overset{\overset{\displaystyle O}{\|}}{C}-CH_3$$

coenzyme A   acetic acid   acetyl CoA

The acetyl group $CH_3\overset{\overset{\displaystyle O}{/\!/}}{C}-$, carried by coenzyme A, can be transferred to any acetyl acceptor during cell metabolism. An example of coenzyme A as an acetyl acceptor is shown in the Krebs tricarboxylic acid cycle.

As its main functions the Krebs cycle (Figure 3) oxidizes the acetyl group, $CH_3\overset{\overset{\displaystyle O}{/\!/}}{C}-$, into CO_2 and water and releases energy, most of which is conserved as ATP.

The Krebs cycle begins when acetyl CoA donates the acetyl group to oxalacetic acid, a four-carbon atom dicarboxylic acid, to form citric acid, a six-carbon tricarboxylic acid, releasing coenzyme A to bring in

1.
$$\begin{array}{l} COOH \\ | \\ C{=}O \\ | \\ CH_2 \\ | \\ COOH \end{array} + CoAS\overset{\overset{\displaystyle O}{\|}}{C}CH_3 \longrightarrow CoASH + \begin{array}{l} COOH \\ | \\ CH_2 \\ | \\ HOC{-}COOH \\ | \\ CH_2 \\ | \\ COOH \end{array}$$

oxaloacetic
acid   coenzyme A citric acid

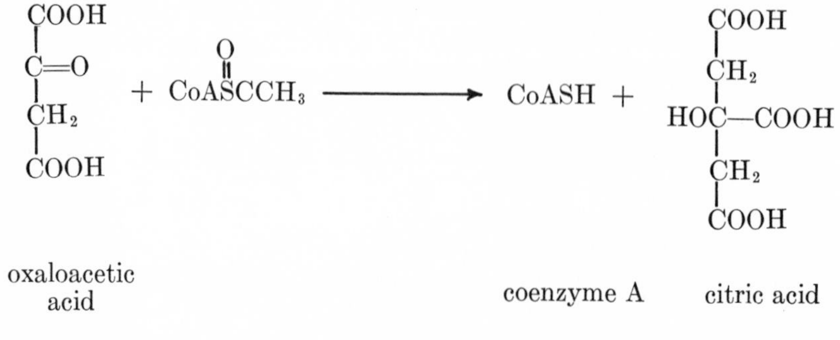

2.

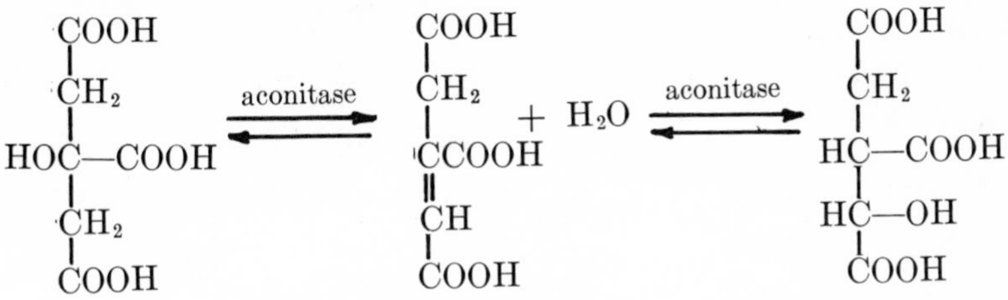

$$
\begin{array}{c}
\text{COOH} \\
| \\
\text{CH}_2 \\
| \\
\text{HOC—COOH} \\
| \\
\text{CH}_2 \\
| \\
\text{COOH}
\end{array}
\quad
\underset{}{\overset{\text{aconitase}}{\rightleftarrows}}
\quad
\begin{array}{c}
\text{COOH} \\
| \\
\text{CH}_2 \\
| \\
\text{CCOOH} \\
\| \\
\text{CH} \\
| \\
\text{COOH}
\end{array}
+ \text{H}_2\text{O}
\quad
\underset{}{\overset{\text{aconitase}}{\rightleftarrows}}
\quad
\begin{array}{c}
\text{COOH} \\
| \\
\text{CH}_2 \\
| \\
\text{HC—COOH} \\
| \\
\text{HC—OH} \\
| \\
\text{COOH}
\end{array}
$$

citric acid   *cis*-aconitic acid   isocitric acid

3.

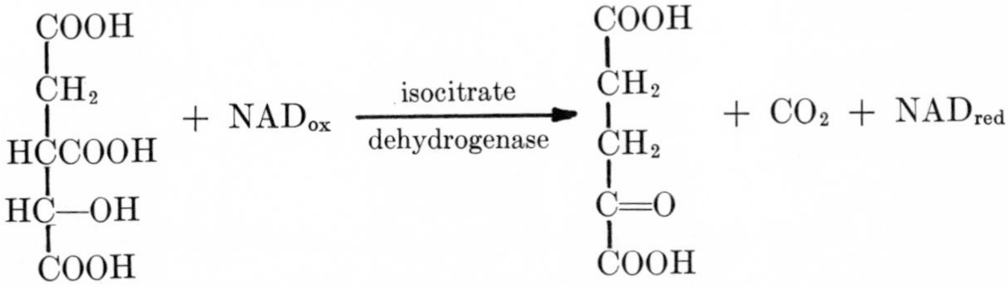

$$
\begin{array}{c}
\text{COOH} \\
| \\
\text{CH}_2 \\
| \\
\text{HCCOOH} \\
| \\
\text{HC—OH} \\
| \\
\text{COOH}
\end{array}
+ \text{NAD}_{ox}
\quad
\xrightarrow[\text{dehydrogenase}]{\text{isocitrate}}
\quad
\begin{array}{c}
\text{COOH} \\
| \\
\text{CH}_2 \\
| \\
\text{CH}_2 \\
| \\
\text{C=O} \\
| \\
\text{COOH}
\end{array}
+ \text{CO}_2 + \text{NAD}_{red}
$$

isocitric acid   α-ketoglutaric acid

4.

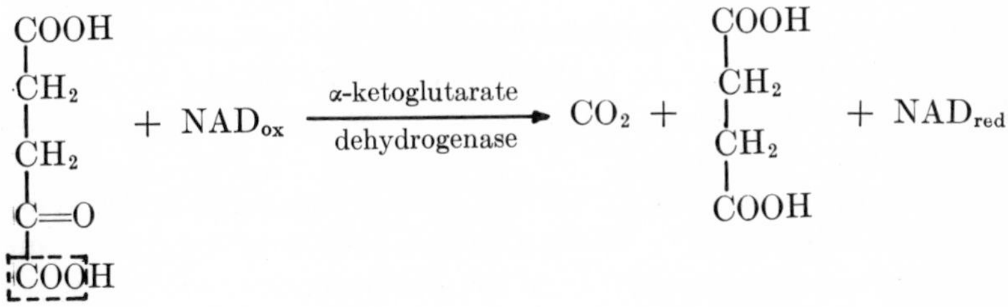

$$
\begin{array}{c}
\text{COOH} \\
| \\
\text{CH}_2 \\
| \\
\text{CH}_2 \\
| \\
\text{C=O} \\
| \\
\overline{\text{COOH}}
\end{array}
+ \text{NAD}_{ox}
\quad
\xrightarrow[\text{dehydrogenase}]{\text{α-ketoglutarate}}
\quad
\text{CO}_2 +
\begin{array}{c}
\text{COOH} \\
| \\
\text{CH}_2 \\
| \\
\text{CH}_2 \\
| \\
\text{COOH}
\end{array}
+ \text{NAD}_{red}
$$

α-ketoglutaric acid   succinic acid

5.

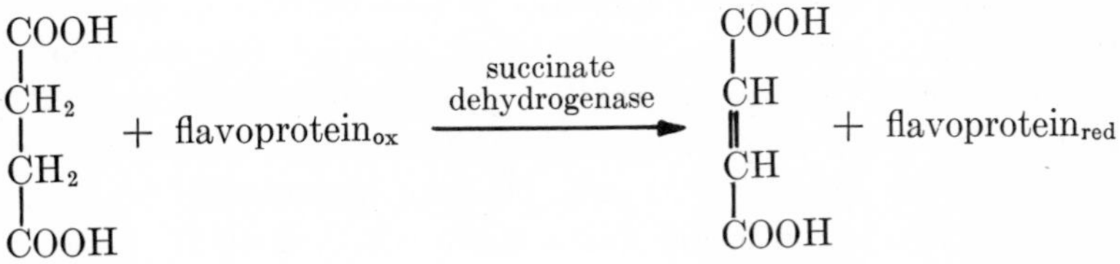

$$
\begin{array}{c}
\text{COOH} \\
| \\
\text{CH}_2 \\
| \\
\text{CH}_2 \\
| \\
\text{COOH}
\end{array}
+ \text{flavoprotein}_{ox}
\quad
\xrightarrow[]{\overset{\text{succinate}}{\text{dehydrogenase}}}
\quad
\begin{array}{c}
\text{COOH} \\
| \\
\text{CH} \\
\| \\
\text{CH} \\
| \\
\text{COOH}
\end{array}
+ \text{flavoprotein}_{red}
$$

succinic acid   fumaric acid

6.
$$
\begin{array}{c}
\text{COOH} \\
| \\
\text{CH} \\
\| \\
\text{CH} \\
| \\
\text{COOH}
\end{array}
+ H_2O
\quad \xrightleftharpoons{\text{fumarase}} \quad
\begin{array}{c}
\text{COOH} \\
| \\
\text{HCOH} \\
| \\
\text{CH}_2 \\
| \\
\text{COOH}
\end{array}
$$

fumaric acid   malic acid

7.
$$
\begin{array}{c}
\text{COOH} \\
| \\
\text{HCOH} \\
| \\
\text{CH}_2 \\
| \\
\text{COOH}
\end{array}
+ NAD_{ox}
\quad \xrightarrow[\text{dehydrogenase}]{\text{malate}} \quad
\begin{array}{c}
\text{COOH} \\
| \\
\text{C}=O \\
| \\
\text{CH}_2 \\
| \\
\text{COOH}
\end{array}
+ NAD_{red}
$$

malic acid   oxaloacetic acid

another acetyl group for the next cycle. The individual steps in the cycle are shown in Equations 1 through 7, and each step is catalyzed by a specific enzyme. It is not necessary to analyze all the equations for the following discussion. The important point is that when a compound is oxidized, it gives up electrons (electron donor), and when a compound is reduced, it accepts electrons (electron acceptor). An oxidation–reduction reaction, therefore, involves a transfer of electrons from one element or one compound to another. Each electron donor tends to give up electrons, and each electron acceptor tends to accept electrons. In other words, some compounds are very hard to oxidize or reduce while others are very easy.

A characteristic affinity for electrons is measured in terms of electromotive force (emf), called standard oxidation–reduction potential; emf values can be arranged from the most negative to the most positive. The most negative (the easiest to lose electrons) can lose electrons to those compounds lower or less negative in the series. Every time electrons are transferred down the series, energy is liberated. The amount depends on the difference in the standard oxidation–reduction potential (emf) between the two compounds. In this respect it is analogous to the biophosphorus compounds listed earlier, in which the compound highest on the list with the greatest free energy of hydrolysis can transfer a phosphate group to form a compound lower in the series. As shown previously, energy is also liberated with each phosphate transfer.

When an acetyl group in the Krebs cycle is oxidized to CO_2 and H_2O, four oxidation or dehydrogenation steps occur in each cycle (Equations 3, 4, 5, and 7). In these oxidation steps the lost electrons do not go directly from the hydrogen atom to the oxygen atom to form water. Instead, in Equations 3, 4, and 7 the electrons first go to the biophosphate electron carrier, NAD_{ox}, to convert it to NAD_{red}; in Equation 5 the electrons are passed on to a flavoprotein$_{ox}$ to be carried as flavoprotein$_{red}$. In the standard oxidation–reduction series NAD_{red} is the most negative and loses its electrons most easily. Next in the series is the electron carrier, flavoprotein, also a phosphorus-containing compound. Oxygen is the least negative in the series, so much so that it goes over to the positive side in the oxidation–reduction potential scale. In between flavoprotein and oxygen are electron carriers called cytochromes; each has an active group containing an iron atom. In its oxidized form, iron has a valence of 3^+ (Fe^{3+}); it can accept an electron and be reduced to Fe^{2+}.

When a compound in the Krebs cycle is oxidized, it loses a pair of electrons. This electron pair does not go directly to the oxygen atom. It goes first to NAD, then to flavoproteins, and down the series through a couple of cytocromes before it reaches oxygen and reduces it to form water. Many details in this sequence of electron transfers are still not clearly understood, but the principle of the total transfer is known. As a sidelight, cyanide is a deadly poison because it blocks the enzymatic transfer of electrons from cytochrome to oxygen. In other words, it inhibits the last step in the respiratory sequence.

The importance of the oxidation reactions in the Krebs cycle to the biological system, of course, lies in the release of energy. When a pair of electrons travels from NAD_{red} to oxygen, 52,000 calories of energy are released; a large portion of this amount is conserved by the formation of three molecules of ATP from ADP and orthophosphate. As you will recall, each ATP formation conserves 7000 calories of energy. What would happen if a poison were introduced which permits the transport of electrons but which interferes with the formation of ATP from ADP and orthophosphate? When such a poison, 2,4-dinitrophenol:

$$O_2N \underset{}{\bigcirc}\!\!\!\begin{array}{c} -NO_2 \\ -OH \end{array}$$

is introduced into a rat, the respiration rate of the rat is increased, accompanied by a rapid increase in body temperature. Since the oxidation energy is not conserved as ATP, it is lost as heat.

Energy is released when glucose is broken down by anaerobic cells. Energy is also released from the aerobic cell oxidation of the acetyl group from the foods we consume *via* the Krebs cycle mechanism. From the standpoint of phosphorus chemistry, a good portion of the released energy is conserved by the formation of ATP from ADP and orthophosphate. When the body needs energy to perform a function, such as flexing a muscle, the energy is supplied by the breakdown of ATP to ADP and orthophosphate.

The roles of other biophosphates, such as NAD which serves as an electron carrier and coenzyme A which serves as an acetyl carrier, have also been described. Many additional phosphorus compounds, indispensable to the functioning of the body, are known, and no doubt more remain to be discovered. The 1971 Nobel prize in physiology and medicine was awarded to Earl W. Sutherland of Vanderbilt University at Nashville, Tenn., for his discovery of cyclic adenylic acid (AMP) and adenyl cyclase, the enzyme that synthesizes cyclic AMP from ATP:

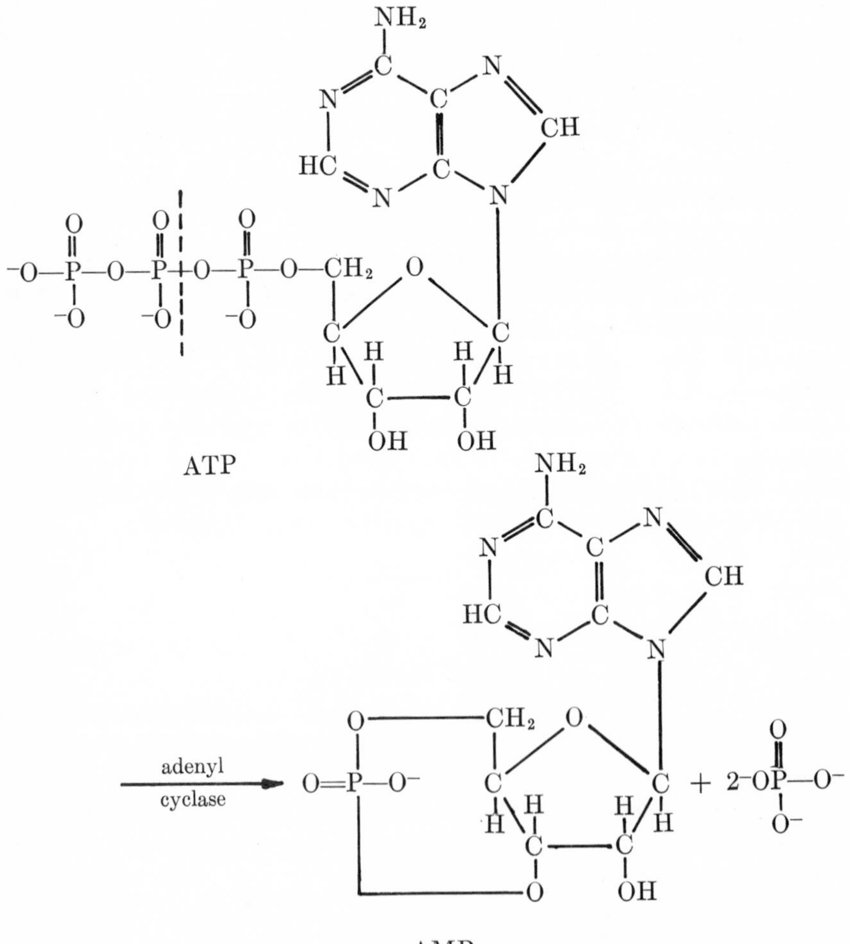

ATP

AMP

This unique phosphorus-containing molecule mediates the action of many hormones and regulates important activities in almost every animal cell. The elucidation of the AMP structure is an interesting example of cooperative efforts among scientists and their alertness in coordinating seemingly unrelated facts. Since AMP is present in cells in extremely small amounts, it was difficult to accumulate enough of it for analysis and characterization. Finally Professor Sutherland had enough evidence that it was a nucleotide and sent his results to Leon Heppel, then at the National Institutes of Health. Dr. Heppel had previously received a letter from David Lipkin of Washington University, St. Louis, describing a nucleotide which he had prepared chemically by treating ATP with barium hydroxide, a base. One day while clearing his desk, Dr. Heppel ran across both letters and recognized that both scientists were probably working on the same compound. He got the two to communicate with each other, and from their cooperative effort the chemical structure of AMP was established.

Reference

1. Albert L. Lehninger, "Bioenergetics," W. A. Benjamin, Inc., New York, 1965.

17

Organic Phosphorus Nerve Gases and Insecticides

You have heard of nerve gases so potent that a few drops will kill a horse and so dangerous that no country dares to use it first in actual warfare. You have benefited indirectly from organic phosphorus insecticides because they increase the yield of agricultural products and reduce the suffering and destruction caused by insect pests. You may be surprised to learn, however, that the reseach which led to the development of organic phosphorus insecticides was carried out in parallel with the basic research that led to the development of nerve gases intended for killing people.

The story of this research illustrates some basic scientific methods. The first thing research chemists must consider is whether or not someone else is thinking and working along a similar or even the exact same line. Even though the original premise formulated for the problem by the two groups may or may not be the same, the chances are good that similar, perhaps identical, results will be obtained.

The second consideration is that most successful compounds are a result of systematic search sparked by some early clue. This search is accompanied by deductive reasoning based on known theories as well as hypotheses developed as the problem progresses. The active lead may be discovered accidentally in one's own laboratory, or it may be inspired by literature on related subjects.

Organic phosphorus poison research programs were carried out during World War II both in England and Germany. In England the program was led by B. C. Saunders at Cambridge University. Gerhard Schrader of the I. G. Farben Works at Elberfeld directed the German program. Since the two countries were at war with each other, neither group supposedly knew what the other was doing.

The Cambridge program aimed to develop a potent chemical warfare agent. In a lecture on the subject, Dr. Saunders stated facetiously that his original reason for picking phosphorus–fluorine compounds was that phosphorus is known to be very poisonous. Thus, a combination of phosphorus and fluorine, another known dangerous chemical, should yield some very poisonous compounds. Of course, such a premise is not always borne out by facts. For example, chlorine is a known poisonous gas, and sodium metal burns explosively in water; yet their combination, sodium chloride, is harmless table salt.

Nevertheless, some premise must be formulated at the beginning of all chemical research, and such a premise is usually based on the best available knowledge. If the original premise is proved wrong after a few experiments, it can be discarded; thus, knowledge is gained even from negative results. New premises are then formulated and evaluated until a clue is discovered.

UPI Photo

A U. S. Army, Rocky Mountain Arsenal technician walks through a shed filled with metal containers of nerve gas. Wearing protective clothing and gas masks, he checks for leaks with a live white rabbit. The Arsenal was opened to the press in February 1970 to show how decontamination of the chemicals will be done.

Dr. Saunders did indeed find that some organic phosphorus fluorine compounds were quite toxic, especially the dialkyl phosphorofluoridates

$$O$$

with the general formula $(RO)_2PF$. Here was the lead he was looking for. A literature search showed that in 1932 in Germany, Lange and his assistant Kruger had reported the synthesis and toxic properties of the

$$O$$

diethyl phosphorofluoridate, $(C_2H_5O)_2PF$. Since Lange and Kruger had other research goals in mind, the toxicity angle was not followed, but now Dr. Saunders' research group pursued this lead assiduously. They synthesized and evaluated many dialkyl phosphorofluoridate derivatives.

They prepared substances in which R groups in the $(RO)_2\overset{\displaystyle O}{\overset{\|}{P}}F$ formula were different. They made compounds in which the fluorine atom was attached directly to the phosphorus atom. They made other compounds in which the fluorine atoms were located on various carbons in the R group. Some of these compounds were toxic, and some were not. Toxicity was first evaluated through animal tests.

For some compounds, Dr. Saunders even used himself as the guinea pig. Several times he almost died after entering a chamber which he thought contained a sub-lethal dose of a toxic organic phosphorus compound. At one stage of his research the British government questioned their continued support of his project. Saunders invited the project officer to a session in the test chamber. When the officer discovered first hand how powerful the agents were, Dr. Saunders had no further difficulty with continued support.

After synthesizing and evaluating many organic phosphorus fluorine compounds of various structures, the researchers chose diisopropyl phosphorofluoridate:

$$(i\text{-}C_3H_7O)_2\overset{\displaystyle O}{\overset{\|}{P}}F)$$

commonly called DFP, as the compound with the correct balance of properties for a chemical warfare agent. Doubtless some of Dr. Saunders' other compounds were not made public because of military secrecy. Some, however, were promoted as agricultural insecticides.

After a compound's effectiveness is determined, a practical process must be developed for its large-scale, economical manufacture. Several are possible. The process which Dr. Saunders' group developed was based on the use of phosphorus trichloride (PCl_3) as the intermediate. One mole of phosphorus trichloride reacts with three moles of isopropyl alcohol ($i\text{-}C_3H_7OH$) to form diisopropyl phosphonate:

$$PCl_3 + 3\ i\text{-}C_3H_7OH \rightarrow (i\text{-}C_3H_7O)_2\overset{\displaystyle O}{\overset{\|}{P}}H + i\text{-}C_3H_7Cl + 2\ HCl$$

diisopropyl phosphonate

Diisopropyl phosphonate then reacts with elemental chlorine (Cl_2) to form diisopropyl phosphorochloridate:

$$
\underset{(i\text{-}C_3H_7O)_2PH}{\overset{O}{\|}} + Cl_2 \rightarrow \underset{\substack{(i\text{-}C_3H_7O)_2PCl \\ \text{diisopropyl} \\ \text{phosphoro-} \\ \text{chloridate}}}{\overset{O}{\|}} + HCl
$$

Diisopropyl phosphorochloridate can be converted to diisopropyl phosphorofluoridate (DFP) by treatment with sodium fluoride (NaF), in a solvent such as benzene:

$$
\underset{(i\text{-}C_3H_7O)_2PCl}{\overset{O}{\|}} + NaF \xrightarrow[\text{solvent}]{\text{benzene}} \underset{(i\text{-}C_3H_7O)_2PF}{\overset{O}{\|}} + NaCl
$$

DFP is extremely toxic. I have seen animals die violently after an infinitesimal dose of this compound is injected. To be exact, the lethal dose needed to kill 50% (LD_{50}) of the rats exposed for 10 minutes is 0.36 ppm (one third of one part per million) in air.

In the United States (and, presumably, other nations) the chemical warfare services have also done extensive research on nerve gases based on organophosphorus compounds. Although large quantities of these compounds have been in storage for use if and when necessary, news reports in 1971 indicated that the government was destroying some of these stores. The exact chemical nature of these organophosphorus gases has not been made public.

In Germany, organophosphorus compounds were studied by Gerhard Schrader to develop an effective insecticide. Consequently many extremely toxic compounds were discovered and diverted for use as chemical warfare agents. Here, the borderline between research on insecticides and on chemical warfare agents became obscured indeed.

Schrader gives an interesting account of how he happened to be working on organic phosphorus compounds as insecticides (*1*). Unlike Saunders, he did not begin by speculating that phosphorus–fluorine compounds might be toxic. He arrived at the toxic organic phosphorus compounds through years of systematic research. His research program was initiated in 1934 when the director of research of the German I. G. Farben factory in Leverkusen instructed him to seek a new type of insecticide because of undesirable properties in insecticides then in use.

At that time inorganic researchers of the I. G. Farben laboratory were working on cheaper methods for preparing fluorine and its compounds. Dr. Schrader decided, therefore, to undertake the systematic synthesis of fluorine-containing compounds. This program was done in collaboration with biologists at Leverkusen who tested the compounds on various insects. Before long they found a fluorine–sulfur compound—methanesulfonyl fluoride (CH_3SO_2F)—that was an effective fumigant against insect pests in stored grains. Unfortunately, in large-scale evaluation it was found to absorb on the grain, which thus became too toxic for human consumption. Researchers then evaluated the poisoned grain as bait for mice; however, the absorbed methanesulfonyl fluoride evaporated gradually from the grain, and toxicity thus decreased with time and finally disappeared.

USDA Photo

Insecticides mean the difference between profit and loss. Untreated cotton, left, yielded one-tenth bale per acre; treated, a bale per acre.

Systematic study of organic fluorine–sulfur compounds continued. Organic fluorine compounds containing a sulfur–nitrogen bond, sulfur–carbon bond, and sulfur–oxygen bond were synthesized. Toxicity was discovered in many of these compounds, and some were almost good enough for commercial insecticide exploitation. Dr. Schrader stated that he finally exhausted the possibilities of preparing fluorine compounds with sulfur as the central atom. In the periodic table of chemical elements phosphorus is next to sulfur. Since these neighbors have related properties, Schrader thought that organic fluorine compounds containing phosphorus as the central atom might also be biologically toxic. The first compound he synthesized in this new series which contained a phosphorus–fluorine bond showed some insecticidal activity. This compound

was bis(N,N-dimethyl)phosphorodiamidic fluoride, $[(CH_3)_2N]_2\overset{\overset{\displaystyle O}{\|}}{P}F$.

This new lead was promising. More compounds were synthesized. Each new structural modification was based on the results of the biological evaluation of the previous series of compounds. Dr. Schrader's systematic research program based on phosphorus as the central atom was extremely fruitful. It led eventually to the development of organic phosphorus compounds suitable as commercial insecticides which contained no fluorine atoms at all. Some of the very potent compounds were diverted to military use.

Among the more important insecticides discovered in Dr. Schrader's laboratory and introduced commercially immediately after World War II were the following:

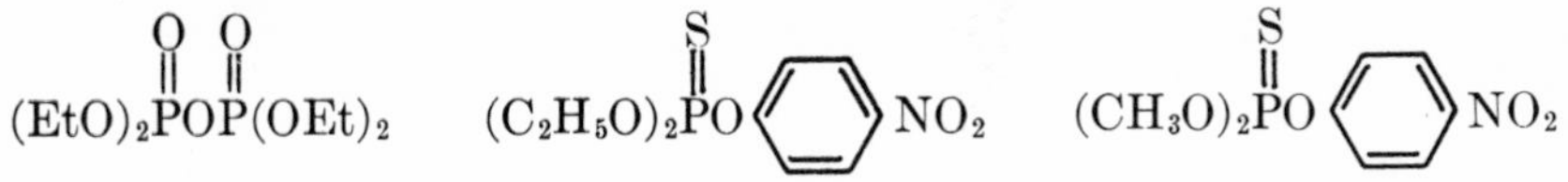

$(EtO)_2\overset{\overset{\displaystyle O}{\|}}{P}O\overset{\overset{\displaystyle O}{\|}}{P}(OEt)_2$ $(C_2H_5O)_2\overset{\overset{\displaystyle S}{\|}}{P}O\text{—}\langle\bigcirc\rangle\text{—}NO_2$ $(CH_3O)_2\overset{\overset{\displaystyle S}{\|}}{P}O\text{—}\langle\bigcirc\rangle\text{—}NO_2$

tetraethyl   ethyl   methyl

pyrophosphate   parathion   parathion

At the end of World War II, when the Americans learned what the Germans had done, various industrial chemical research laboratories began their own research to discover even more effective insecticides. There was no doubt that Schrader's compounds were very effective; however, they were also very toxic to warm-blooded animals, including humans. Because early manufacturers were unfamiliar with these compounds, several workers died from accidental poisoning. In addition many users of these insecticides on farms and in fruit orchards did not quite believe how toxic these compounds were and did not follow the labeled precautions. Several deaths resulted.

Although the American researchers were trying to develop compounds which would be highly toxic to insects but relatively non-toxic to humans and animals, theoretical knowledge was insufficient to guide such a program at that time. Most programs just followed Schrader's systematic approach. In other words, a series of new structures were synthesized. They were tested against standard insects. If a compound showed some activity, its structure was varied to determine which substituent would enhance or decrease its desirable properties. Statistically, of an average of about 3000 compounds synthesized, one showed commercial possibilities.

At Hercules Agricultral Chemicals Laboratory, various types of insect pests are released on plants treated with new compounds displaying insecticidal activity. Plants are later examined for extent of insect damage.

While these new synthesis programs were going on, other research laboratories in the United States were evaluating the organophosphorus compounds already on their shelves. In most cases these compounds had been synthesized for other purposes—*e.g.*, as flotation agents for separating metal ores from sand and clay, lubricating-oil additives, and plasticizers. Candidates which had failed were therefore included in the insecticide screening program. Two outstanding organophosphorus insecticides were discovered by this approach. One is the well-known, malathion:

$$(CH_3O)_2\overset{\overset{\text{S}}{\|}}{P}-S-\overset{\overset{\displaystyle CH_2\overset{\overset{\text{O}}{\|}}{C}-OEt}{|}}{CH}-\overset{}{C}-OEt)$$

I was told that this compound is related to a class of compounds originally synthesized for testing as flotation agents. It is a remarkable compound because it is toxic to a wide range of insects and yet is absolutely harmless to mammals. It is, therefore, ideal for use in home gardens. The rotten-egg odor encountered in most insect dusts used in rose gardens is usually the result of malathion in the formulation.

Another compound which was a relative of one originally synthesized as a lubricating oil additive has the formula:

$$\overset{\text{S}}{\underset{\|}{}}$$
$$(C_2H_5O)_2PSCH_2SC_2H_5$$

It is known as Thimet and is an extremely effective systemic insecticide, but is quite toxic to warm-blooded animals.

Another interesting example is $N(\beta$–O,O, diisopropyl dithiophosphorylethyl) benzenesulfonamide:

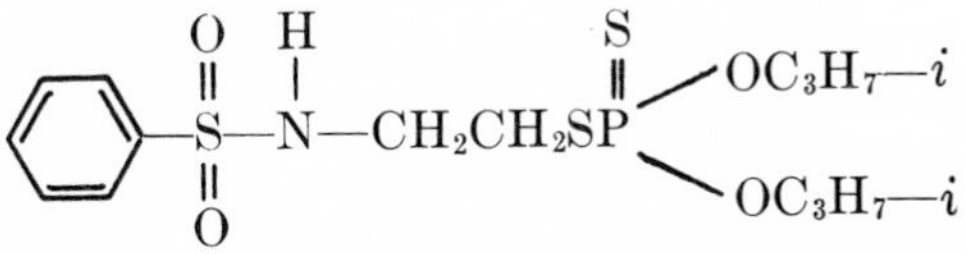

known by the tradename Betasan. It was synthesized during a systematic research program to find effective insecticides. Although it was not an effective insecticide, continued evaluation by a botanist showed it to be a very good herbicide. It effectively and selectively kills crabgrass weed as it emerges from the seed without harming grass and other ornamental lawn or garden plants.

Most of the effective pesticides were not discovered accidentally. If chemists and biologists had depended entirely on luck, they probably wouldn't have discovered even one successful compound out of an average of 3000. The odds would have been much higher. After researchers discover several compounds in a series that show biological activity, they formulate a theory for such activity. Then they design model compounds to check their hypothesis. For the organophosphorus pesticide research, two major questions usually must be answered. One is, what structural type of organic phosphorus compounds is effective—*i.e.*, which functional group or groups in the compound are essential to activity. The second is, what is the mechanism for the toxic action. That is, how do the compounds kill insects and/or warm-blooded animals. Wherever possible, it is important to correlate chemical theory with biological theory. The mechanisms of toxic action of some organic phosphorus compounds are discussed in the next chapter.

Literature Cited

1. Gerhard Schrader, "Die Entwicklung neuer insektizider Phosphorsäure-Ester," Verlag Chemie GmbH, Weinheim/Bergstr., 1963.

18

Mechanism of Toxic Action of Organic Phosphorus Poisons

Why are some organophosphorus compounds such potent killers of insects or animals? How do they work? What are the antidotes, and how were they discovered? At present, not all the answers on the mechanism of biological action of these compounds are known. However, the biological theory developed thus far fits quite well with the structure and chemistry of some of these poisons. The theoretical basis for their action was developed from hindsight. That is, the toxic action was discovered first and the theory that accounts for it was formulated and checked out later.

Nevertheless, the theories, though incomplete, have served as very useful guiding principles in the search for newer and more effective compounds. They also serve as guides for the development of antidotes. However, since we do not have the complete mechanistic picture, a bit of luck still plays a big part in each discovery. Such luck usually goes, however, to researchers who are diligent and alert enough to take advantage of unexpected leads.

Biologists generally agree that a poisonous organic phosphorus compound is toxic because it inhibits the enzyme, acetylcholinesterase (AChE). Although the exact mechanism is still a subject for further research, there are sufficient data to indicate that the following simplified description is valid. In a nerve transmission system there is a small gap between the end of the nerve fiber and the end of the muscle fiber. The gap is called the myoneural junction. It is about 100 A (one hundred-thousandth of a millimeter) wide. An impulse for action in the muscle is started by changes in the electrical potential of the nerve ending. This releases a chemical called acetylcholine, $[CH_3\overset{\displaystyle O}{\overset{\|}{C}}-OCH_2CH_2\overset{+}{N}(CH_3)_3]$, which diffuses across the myoneural junction and is accepted by the muscle fiber. Acetylcholine changes the electric potential of the muscle fiber and causes the fiber to contract. The contraction is what we observe or feel as muscle action.

After acetylcholine contracts the muscle fiber, it is immediately hydrolyzed by the enzyme acetylcholinesterase (AChE). The products

$$O$$
$$\parallel$$

of hydrolysis are acetic acid ($CH_3C{-}OH$) and choline [$HOCH_2CH_2\overset{+}{N}$-(CH_3)$_3$]. These two compounds then diffuse away from the muscle fiber. They will be brought together again by other body enzymes elsewhere.

Acetylcholine must be hydrolyzed and removed from the muscle once it contracts the fiber. When acetylcholine is gone, the muscle fiber returns to its original state and is ready for contraction again when the next controlled impulse from the nerve fiber sends in more acetylcholine. If acetylcholine remained in the muscle fiber longer than usual, the muscle would twitch and remain contracted.

If another nerve impulse sends in more acetylcholine before the acetylcholine from the previous impulse is dissipated, the muscle fiber cannot cope with it and becomes paralyzed. Paralyzed muscles are usually uncomfortable but not fatal unless the chest muscles that control breathing are involved; in that case one dies quickly from asphyxia. Thus, the enzyme acetylcholinesterase must be available to hydrolyze acetylcholine and move it away from the muscle.

According to present theory, certain organic phosphorus compounds are toxic because they inhibit AChE so that it can no longer hydrolyze acetylcholine. This inhibition is chemical. Let's see how it works.

The complete chemical and physical structure of the enzyme AChE is still not known. We do know that it is a large protein molecule made up of many amino acids and exists as a coiled chain. The enzyme has two known active sites. One is called the esteratic site because it attracts ester groups to it. The other is called anionic or negative site because it attracts a cation (or an electrically positive section of a compound). The distance between these two sites is estimated at 4.5A. When AChE acts on acetylcholine, the first step is the formation of a complex. In this complex, the anionic or negative site of the AChE locks onto the cationic or positive end of the acetylcholine molecule. The esteratic site locks

$$O$$
$$\parallel$$

into the acetyl($CH_3C{-}$) end of the acetylcholine.

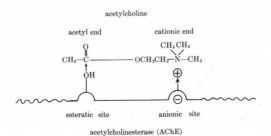

acetylcholine

acetyl end   cationic end

esteratic site   anionic site

acetylcholinesterase (AChE)

In this intimate condition it's easy for the esteratic site to react with the acetyl group $\left(\text{CH}_3\overset{\displaystyle O}{\overset{\|}{\text{C}}}-\right)$ and form acetylated acetylcholinesterase.

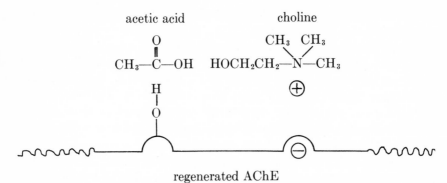

acetylated acetylcholinesterase

This reaction breaks up acetylcholine. Acetylated acetylcholinesterase then hydrolyzes in the water of the surrounding tissue to regenerate AChE and liberate acetic acid.

acetic acid   choline

$$\underset{\text{H}}{\overset{\displaystyle O}{\underset{\displaystyle |}{\overset{\|}{\text{CH}_3-\text{C}-\text{OH}}}}} \qquad \overset{\displaystyle \text{CH}_3 \quad \text{CH}_3}{\underset{\oplus}{\text{HOCH}_2\text{CH}_2-\text{N}-\text{CH}_3}}$$

regenerated AChE

Acetic acid and choline then diffuse away from the muscle fiber and reform for use again in the next nerve impulse.

A crude mixture of AChE enzyme isolated from a live biological system, such as a homogenate of fly brains, can split acetylcholine in a test tube. If the test tube contains a buffer solution of sodium bicarbonate (NaHCO_3), acetic acid $\left(\text{CH}_3\overset{\displaystyle O}{\overset{\|}{\text{C}}}-\text{OH}\right)$ which is liberated reacts with sodium bicarbonate to give off carbon dioxide gas:

$$\text{CH}_3\overset{\displaystyle O}{\overset{\|}{\text{C}}}-\text{OH} + \text{NaHCO}_3 \rightarrow \text{CH}_3\overset{\displaystyle O}{\overset{\|}{\text{C}}}-\text{ONa} + \text{H}_2\text{O} + \text{CO}_2 \uparrow$$

[The above test carried out away from the biological specimen is called *in vitro*, in contrast to an *in vivo* test conducted in the live specimen.] If the AChE is inhibited or prevented from splitting acetylcholine to yield acetic acid, no gaseous CO_2 will evolve. If AChE is partially inhibited, less CO_2 is evolved. The biologist uses this *in vitro* test extensively to measure the degree of AChE inhibition by agents such as organic phosphates.

Two sites in AChE must be locked onto two sites in the acetylcholine before AChE can be acetylated. This is proved by the fact that when the anionic site in AChE is occupied by a stronger cation from another molecule, AChE can no longer hydrolyze acetylcholine.

Present theory for organic phosphorus poisoning proposes that an organophosphorus molecule (an ester) forms an intimate complex with AChE. AChE is thus phosphorylated, rather than acetylated. For example, with diethyl *p*-nitrophenyl phosphate:

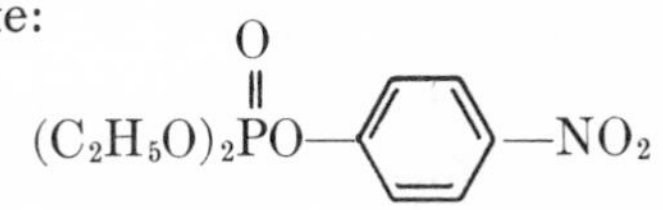

or para-oxon, the phosphorylated AChE is the diethyl phosphate ester of AChE. [I use para-oxon as an example because the widely used parathion:

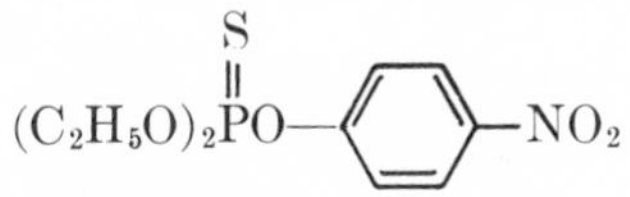

has been proved experimentally to be converted to para-oxon in the body before it becomes toxic.] Most researchers agree that the esteratic site of AChE contains an –OH group from the amino acid serine. The serine –OH is therefore the site for phosphorylation:

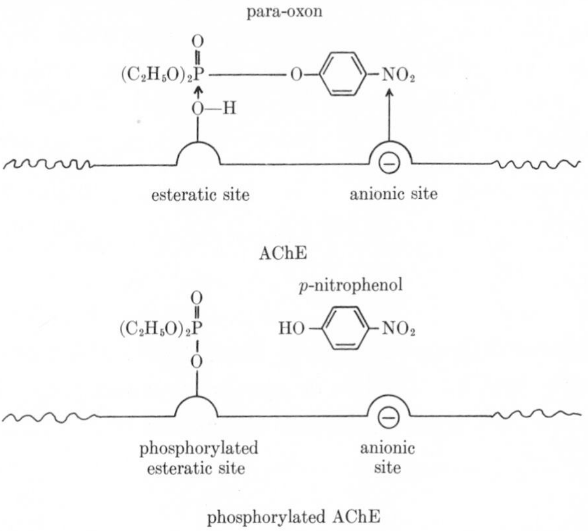

para-oxon

esteratic site   anionic site

AChE

p-nitrophenol

phosphorylated esteratic site   anionic site

phosphorylated AChE

The diethyl phosphate group $[(C_2H_5O)_2\overset{\overset{O}{\|}}{P}$—$]$ attached to the esteratic site of the AChE makes it unavailable for forming a complex with incoming acetylcholine. AChE, thus cannot split acetylcholine, and we say that the AChE is inhibited. When excess unsplit acetylcholine accumulates on the muscle, the muscles fiber starts to twitch and is finally paralyzed.

The diethyl phosphate group in phosphorylated AChE is not hydrolyzed off as is the acetyl group in acetylated AChE. However, hydrolysis does occur, slowly, to regenerate AChE so that it becomes available again for splitting acetylcholine. For this reason if you aren't killed quickly by an organic phosphorus poison, you'll live. In other words, if the AChE is not inhibited by an overdose of organic phosphorus poison, the inhibited AChE may be regenerated in time to prevent death.

In the above description of the formation of the complex between AChE and acetylcholine, much was said about the esteratic site complexed with the ester group and the anionic site complexed with the cationic site. In the case of an organic phosphorus compound, such as

para-oxon, a phosphate ester group $(C_2H_5O)_2\overset{\overset{O}{\|}}{P}$— was available to complex with the esteratic site of the AChE. However, we can't really say that there is also a definitive cationic or positive site to complex with the anionic or negative site of the AChE.

Some researchers speculate that complexing may be caused by a spatial effect. That is, the molecular structural geometry of the other portion of the para-oxon molecule fits into the AChE chain like a key in a lock. This molecular fit is of the utmost importance. If, in addition, there also happens to be a cationic site in the phosphorus compound in the correct position to lock onto the anionic site of AChE, it enhances the inhibition of AChE. This last point was shown by the comparison of

the activity of the compound $(C_2H_5O)_2\overset{\overset{O}{\|}}{P}SCH_2CH_2\overset{\oplus}{S}Et_2$ with that of

$(C_2H_5O)_2\overset{\overset{O}{\|}}{P}SCH_2CH_2SEt$. Pictorially these two compounds are structurally similar; however, the first compound containing a cationic (positive) site reacts 1000 times faster than the second which does not possess a cationic site.

Antidotes

What remedies are there for organic phosphorus poisoning?" The therapeutic remedy devised is based on the theories of toxic action discussed above. If the poisoning isn't too severe and is of short duration, two methods can be applied together. They are based on the fact that nerve stimuli send acetylcholine to the muscle fiber and AChE in the muscle fiber, inhibited by phosphorylation, is unable to split acetylcholine.

In one method the muscle fiber tip is covered with a shot of atropine; this blocks the muscle fiber from further acetylcholine invasion, and phosphorylated AChE will gradually lose its phosphate group by slow hydrolysis. Once freed of the phosphate group, AChE then resumes its regular duty of splitting acetylcholine.

However, when a muscle is paralyzed, especially the muscles of breathing, AChE must be regenerated as fast as possible. Here chemists have developed a method using oximes. One of the most promising oxime is 2-hydroxyiminomethylmethylpyridinium iodide, or P2-AM:

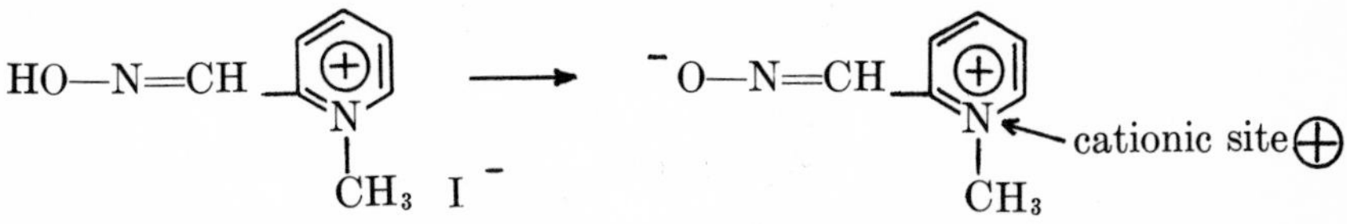

P2-AM has a cationic site that can lock into the anionic site in AChE. The other end of the molecule carries a negative charge that seeks out the positive phosphorus atom in the phosphate ester group and displaces it away from the AChE; this action is called nucleophilic displacement.

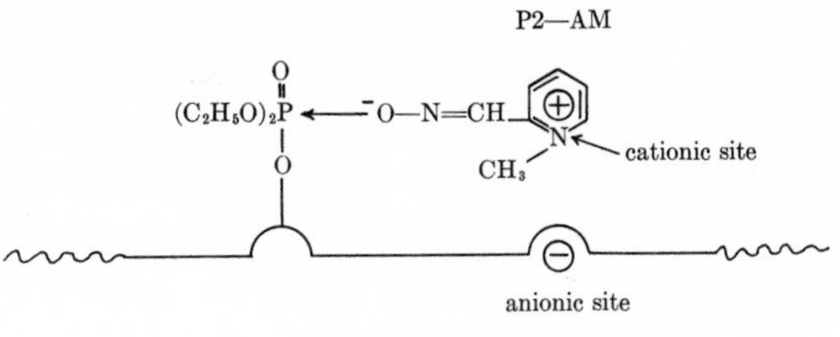

P2—AM

Phosphorylated AChE

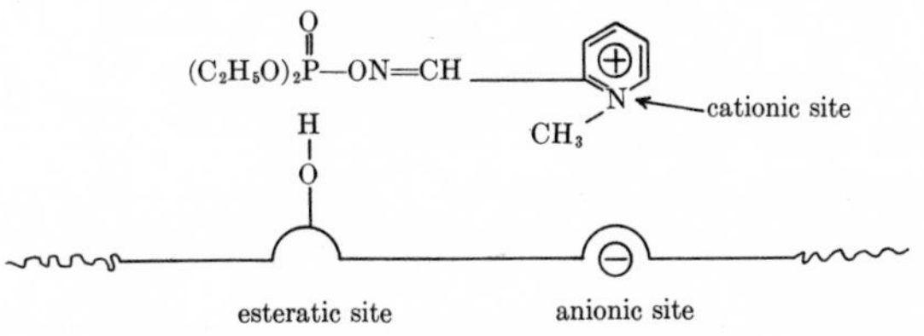

Regenerated AChE

When AChE is regenerated with P2-AM, the compound

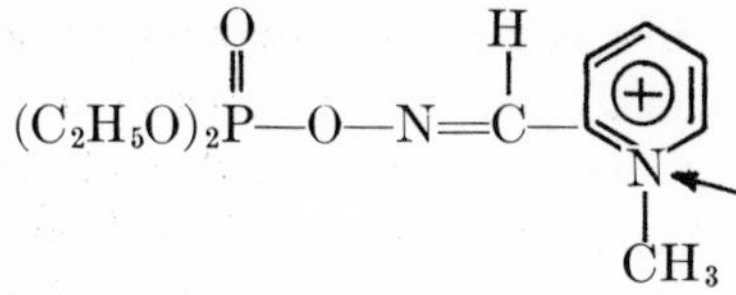

$$(C_2H_5O)_2\overset{\overset{\displaystyle O}{\|}}{P}\!-\!O\!-\!N\!=\!\overset{\overset{\displaystyle H}{|}}{C}\!-\!\underset{\displaystyle CH_3}{\overset{\oplus}{N}}$$

← cationic site

is also produced. This by-product also has a cationic site and a phosphate ester group. Since such a structure should be ideal for phosphorylating the enzyme AChE, it should also be a poison. In other words the above reaction should be reversible, and this is indeed the case. In poisoning by the nerve gas Sarin:

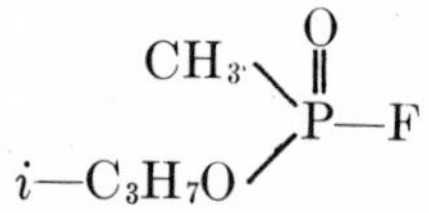

$$\underset{i\!-\!C_3H_7O}{\overset{CH_3}{\diagdown}}\!\!\overset{\overset{\displaystyle O}{\|}}{P}\!-\!F$$

the antidote P2-AM, in displacing the

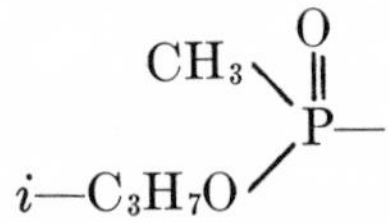

$$\underset{i\!-\!C_3H_7O}{\overset{CH_3}{\diagdown}}\!\!\overset{\overset{\displaystyle O}{\|}}{P}\!-\!$$

group from AChE, forms a new compound which is more toxic than Sarin itself, but the new compound is less stable to hydrolysis than Sarin, so it is destroyed by hydrolysis in the surrounding aqueous medium; this shifts the equilibrium towards AChE regeneration.

Para-oxon reacts with AChE despite the fact that it has no obvious cationic site for AChE to lock onto. Perhaps the molecule fits well geometrically with the enzyme AChE. This speculation on the stereospecific molecular geometry of two compounds fitting together received some support from the experimental evidence with the isomer of P2-AM, the P3-AM. In P2-AM, the six-corner pyridine ring is numbered with the corner occupied by the N as 1, thus:

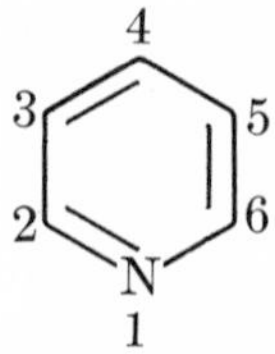

Pyridine ring

The chemical name of 2-hydroxyiminomethyl methylpyridinium iodide means that the hydroxyiminomethyl group, HON=CH—, is attached to the ring on corner 2. In shortening the name to P2-AM, the 2 is retained to indicate the location of the attachment. If the attachment of the HON=CH— group is on corner 3, the compound would be 3-hydroxyiminomethyl methylpyridinium iodide with the structure:

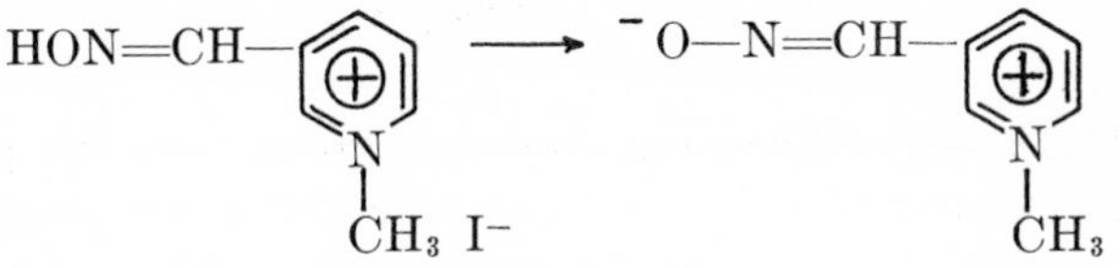

$$HON=CH-\underset{\underset{CH_3 \ I^-}{|}}{\overset{\oplus}{\bigcirc}} \longrightarrow \ ^-O-N=CH-\underset{\underset{CH_3}{|}}{\overset{\oplus}{\bigcirc}}$$

P3—AM

The compound is thus called P3-AM. The substituent group of P2-AM and P3-AM is therefore the same, but its location on the ring is different. P2-AM and P3-AM are isomers of each other. Although P3-AM also has a cationic site, it is not a regenerator for phosphorylated AChE, probably because it does not physically lock with AChE. The presence of the $^-ON=CH-$ group in corner 3 of the ring gives it the wrong spatial configuration.

The longer one waits to use an antidote, the more difficult it is to regenerate AChE. If phosphorylated AChE is allowed to age, AChE regeneration becomes more difficult. On aging, one of the organic groups in the phosphate group is hydrolyzed off. For example, in the inhibition of AChE by diisopropyl fluorophosphate (DFP), the phosphorylated AChE is:

$$i-C_3H_7O \diagdown \overset{O}{\underset{}{\parallel}}$$
$$\underset{i-C_3H_7O \diagup}{\overset{}{P}}\underset{O}{\overset{}{}}$$

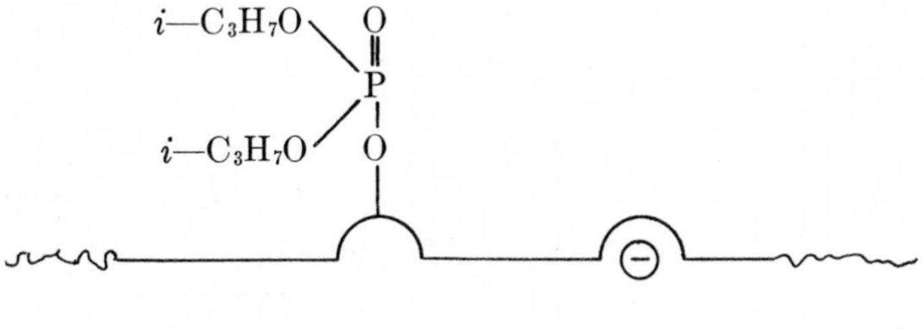

This is a neutral phosphate ester. On aging, one of the isopropyl groups is hydrolyzed off and becomes:

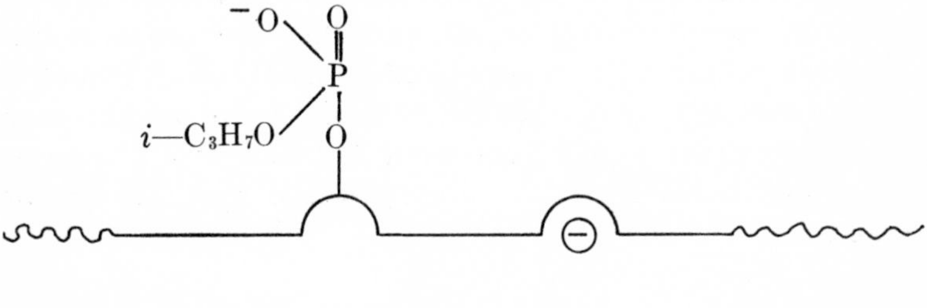

$$^-O \diagdown \overset{O}{\underset{}{\parallel}}$$
$$\underset{i-C_3H_7O \diagup}{\overset{}{P}}\underset{O}{\overset{}{}}$$

Reactivation or regeneration of AChE depends on attack of the phosphorus atom by an anion either an $^-$OH from water or

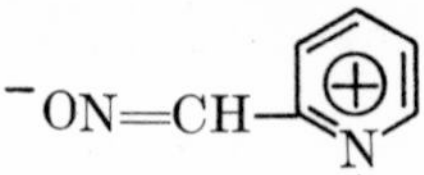

$$^-\mathrm{ON}{=}\mathrm{CH}{-}\overset{\oplus}{\underset{\mathrm{N}}{\bigcirc}}$$

from P2-AM. Aging removes one isopropyl group from the phosphorylated AChE to leave:

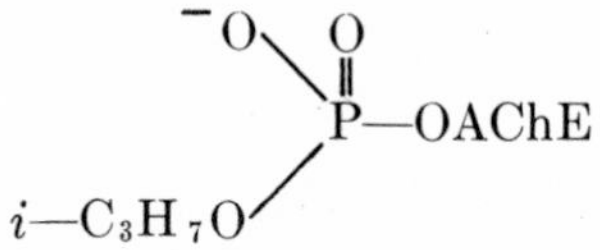

$$\begin{array}{c} \overset{\displaystyle ^-\mathrm{O}}{\underset{i-\mathrm{C_3H_7O}}{\diagdown}}\overset{\displaystyle \mathrm{O}}{\underset{}{\overset{\|}{\mathrm{P}}}}{-}\mathrm{OAChE} \end{array}$$

also an anion. This negatively charged anion will repel any oncoming negatively charged anions. Thus, the effectiveness of:

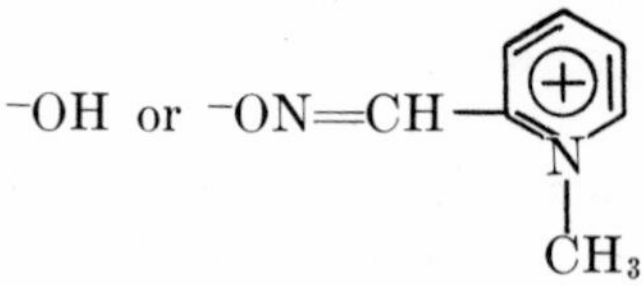

$$^-\mathrm{OH}\ \text{or}\ ^-\mathrm{ON}{=}\mathrm{CH}{-}\overset{\oplus}{\underset{\underset{\mathrm{CH_3}}{\mathrm{N}}}{\bigcirc}}$$

to hydrolyze monoisopropyl phosphate group:

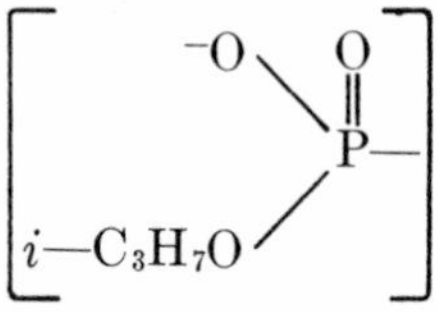

$$\left[\ \overset{\displaystyle ^-\mathrm{O}}{\underset{i-\mathrm{C_3H_7O}}{\diagdown}}\overset{\displaystyle \mathrm{O}}{\overset{\|}{\mathrm{P}}}{-}\ \right]$$

to regenerate AChE becomes very difficult. As a result the patient dies.

If we know a theory for the poisoning action of organic phosphorus compounds and also a theory for counteracting this toxic action, we ought to be able to design compounds of specific structures to get desired toxic activities. To a limited extent, this is now being done. At present the theories are only rough. We still don't know enough about the physiology of mammals and insects. For that matter, we know very little about any one insect. Furthermore, the AChE's are different in humans, in animals, and in insects. Even in insects the enzymes differ from species to species.

Mammals contain other enzymes besides AChE which hydrolyze all phosphate esters and aren't inhibited by them. There are also enzymes which are quite similar to AChE and which, for the lack of a better name, are called pseudocholinesterases. These enzymes hydrolyze practically everything.

Theory has guided us to a point where we can predict that some organic phosphorus compounds of a certain structure will be toxic to mammals and insects. However, the ability to synthesize such compounds is not enough for modern industrial research since there are already successful compounds on the market which have fulfilled most of the requirements. Industrial research must find compounds which are either more effective against a wider range of insects, or cheaper, or both.

Another alternative is to discover compounds that have unique properties not present in existing compounds. Lastly, there is always a place for more economical methods for preparing existing compounds. In other words, the frontier is still wide open for research in the area of new, low-cost, effective insecticides which are selective against harmful pests but not against beneficial insects and animals.

19

Some Interesting Toxic Organic Phosphorus Compounds

I have had some experience in research on organic phosphorus compounds that were possible insecticides. When I first heard that tetraethyl pyrophosphate, TEPP, $\left[(C_2H_5O)_2\overset{\overset{O}{\|}}{P}O\overset{\overset{O}{\|}}{P}(OC_2H_5)_2 \right]$, was effective as an insecticide, I looked through the literature for known methods of preparation. If there was a practical way to make it in the pure form, I wanted to adapt it to large-scale industrial production. I found none, but laboratory research resulted in a practical and economical process. Unfortunately, the use of pure TEPP as an insecticide did not develop into a very large business, and no large plant was ever needed. However, I used the new methods to synthesize analogs of TEPP such as tetrapropyl pyrophosphate $\left[(C_3H_7O)_2\overset{\overset{O}{\|}}{P}O\overset{\overset{O}{\|}}{P}(OC_3H_7)_2 \right]$ and tetrabutyl pyrophosphate $\left[(C_4H_9O)_2\overset{\overset{O}{\|}}{P}O\overset{\overset{O}{\|}}{P}(OC_4H_9)_2 \right]$. These compounds were less toxic than TEPP to mammals, but unfortunately they were also less toxic to insects.

Using the same chemical principle as in that process, I developed a process for synthesizing a sulfur analog of TEPP, tetraethyl dithionopyrophosphate, thio-TEPP: $(C_2H_5O)_2\overset{\overset{S}{\|}}{P}O\overset{\overset{S}{\|}}{P}(OC_2H_5)_2$. This compound was also quite toxic to mammals. Schrader's work showed that he too had made the sulfur analog of TEPP but by a different method. Since I had a pretty good process for thio-TEPP, I decided to synthesize some tetrapropyl dithionopyrophosphate: $\left[(C_3H_7O)_2\overset{\overset{S}{\|}}{P}O\overset{\overset{S}{\|}}{P}(OC_3H_7)_2 \right]$. Because of the results obtained with the oxygen analogs, this compound was expected to be less toxic than thio-TEPP to warm-blooded animals and also less toxic to insects. In other words, I didn't expect much. To my surprise, I found that for practical purposes this compound is not toxic to warm-blooded animals at all but is quite toxic to several species of insects. In fact, it is extremely effective against the chinch bug, a pest that destroys lawns in the southern states along the eastern seaboard. This compound is now known as Aspon.

I have also discovered a compound with the formula:

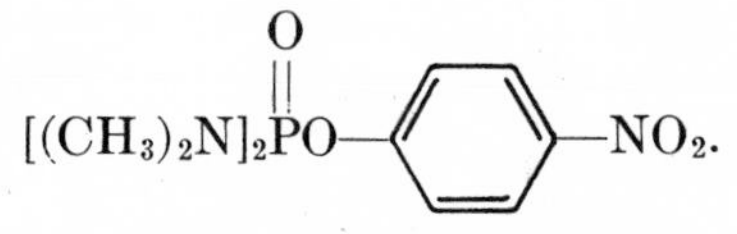

$$[(CH_3)_2N]_2 \overset{\overset{\displaystyle O}{\|}}{P} O \text{—} \bigcirc \text{—} NO_2.$$

It does not harm most insects, but it is quite toxic to warm-blooded animals. I can find no theoretical explanation for this, and the compound seems to have no practical use. Table I lists some of the representative examples of organic phosphorus poisons.

Table I. Representative Organic Phosphorus Poisons

Nerve Gases

Common Names	Formula	Chemical Names
Tabun	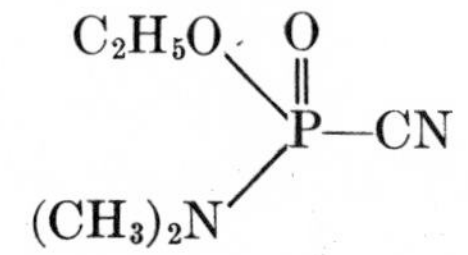	O-ethyl N-dimethyl-phosphoramido-cyanidate
Soman	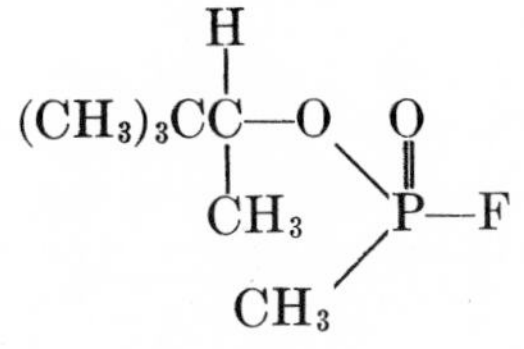	1-methyl-2,2-dimethylpropyl methylphosphono-fluoridate
Sarin	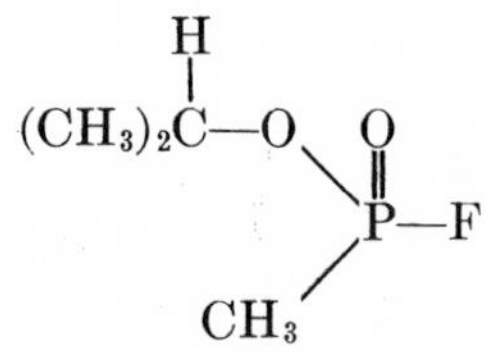	O-isopropyl methylphosphono-fluoridate
DFP	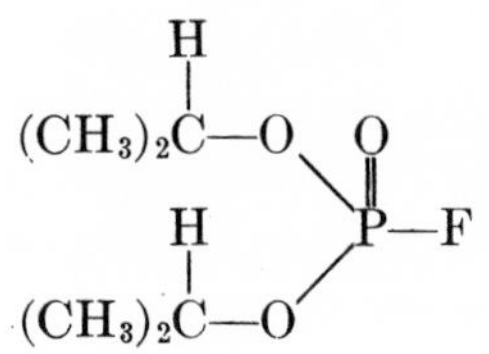	O,O-diisopropyl phosphoro-fluoridate

Table I.   Continued

Contact Insecticides

Common and Trade Names	Formula	Chemical Names
methyl parathion	$(CH_3O)_2\overset{\displaystyle S}{\overset{\|}{P}}O$—⟨benzene⟩—$NO_2$	O,O-dimethyl O-p-nitrophenyl phosphorothioate
parathion	$(C_2H_5O)_2\overset{\displaystyle S}{\overset{\|}{P}}O$—⟨benzene⟩—$NO_2$	O,O-diethyl O-p-nitrophenyl phosphorothioate
Sumithion	$(CH_3O)_2\overset{\displaystyle S}{\overset{\|}{P}}$—O—⟨benzene, CH_3⟩—NO_2	O,O-dimethyl O-(4-nitro-m-tolyl) phosphorothioate
EPN	$\begin{matrix} C_2H_5O \\ \\ C_6H_5 \end{matrix}\!\!\overset{\displaystyle S}{\overset{\|}{P}}O$—⟨benzene⟩—$NO_2$	O-ethyl O-p-nitrophenyl phenylphosphono-thioate
Guthion	$\begin{matrix} CH_3O \\ \\ CH_3O \end{matrix}\!\!\overset{\displaystyle S}{\overset{\|}{P}}SCH_2N$ ⟨benzotriazinone ring system⟩	O,O-dimethyl S-4-oxo-1,2,3-benzotriazin-3(4H)-ylmethyl phosphorodithioate
malathion	$\begin{matrix} CH_3O \\ \\ CH_3O \end{matrix}\!\!\overset{\displaystyle S}{\overset{\|}{P}}SCH$—$\overset{\displaystyle O}{\overset{\|}{C}}$—$OC_2H_5$ $\|$ $CH_2\overset{\displaystyle O}{\overset{\|}{C}}$—$OC_2H_5$	diethyl mercapto succinate, S-ester with O,O-dimethyl phosphorodithioate
Trithion	$\begin{matrix} C_2H_5O \\ \\ C_2H_5O \end{matrix}\!\!\overset{\displaystyle S}{\overset{\|}{P}}SCH_2S$—⟨benzene⟩—$Cl$	S-[(p-chlorophenyl-thio)methyl] O,O-diethyl phosphorodithioate

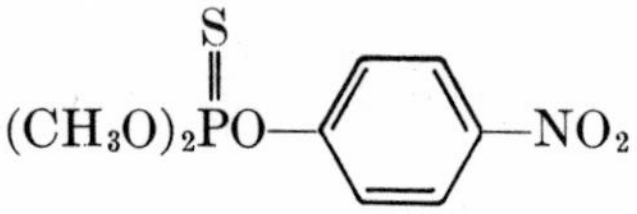

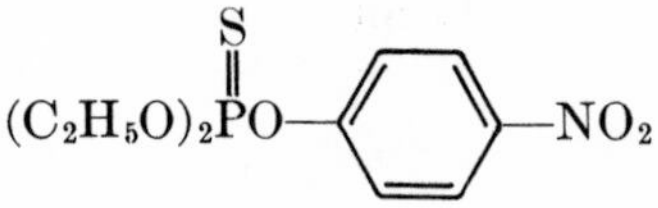

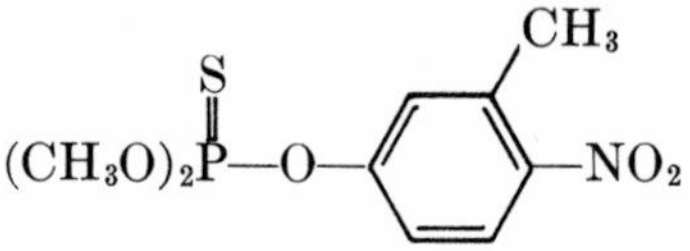

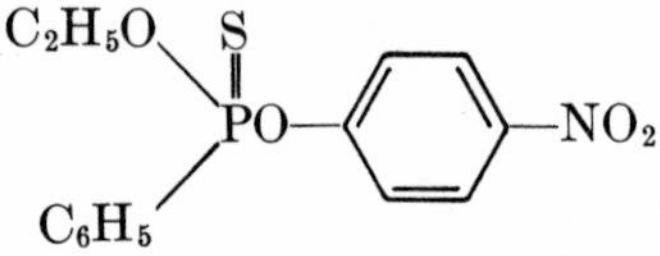

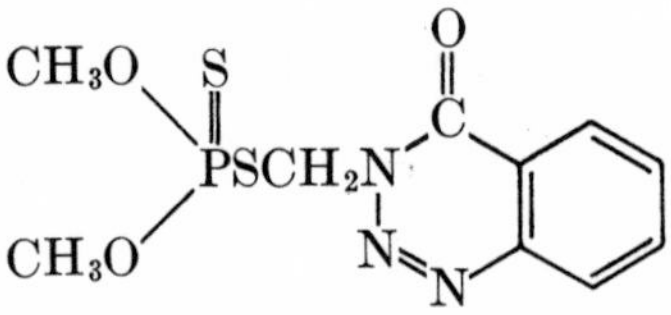

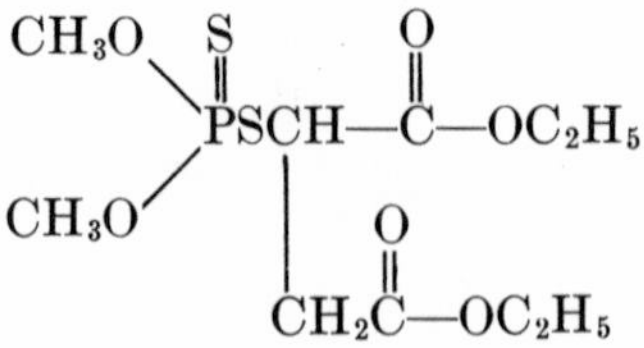

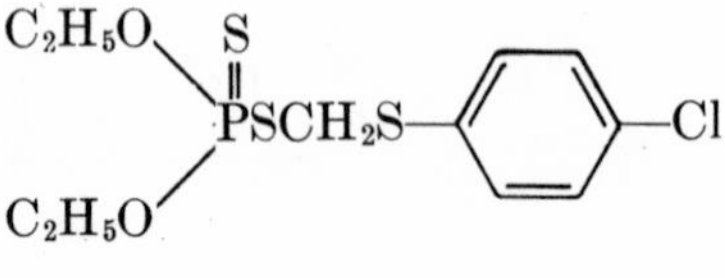

Table I.   Continued

Contact Insecticides

Common and Trade Names	Formula	Chemical Names
Dyfonate	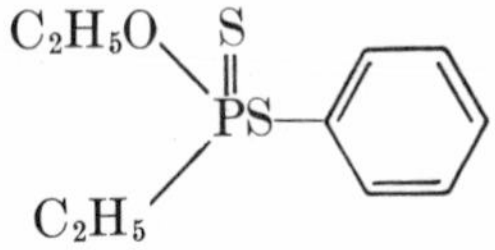	*O*-ethyl *S*-phenyl ethyl-phosphonodithioate
Diazinon	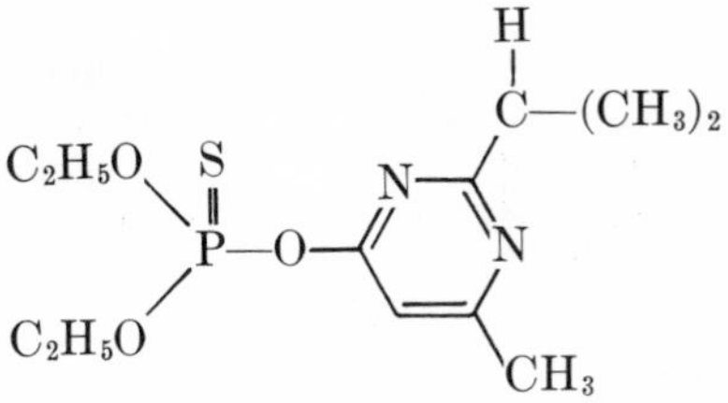	*O,O*-diethyl *O*-(2-isopropyl-6-methyl-4-pyrimidinyl) phosphorothioate
phosphamidon	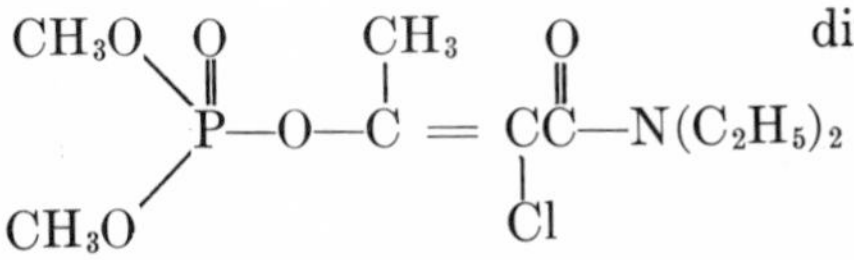	dimethyl phosphate, ester with 2-chloro-*N,N*-diethyl-3-hydroxycrotonamide
ethion	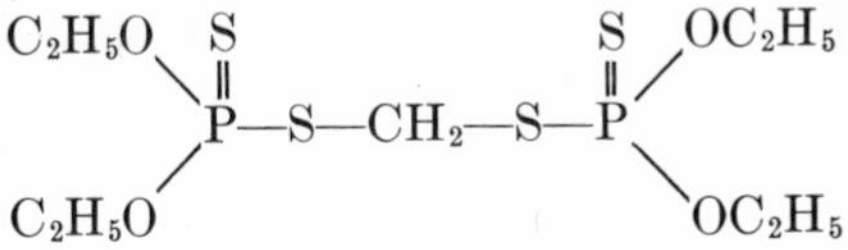	*O,O,O',O'*-tetraethyl *S,S'*-methylene bis-phosphorodithioate
Dipterex	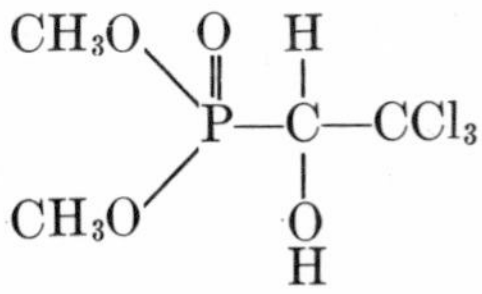	dimethyl (2,2,2,-trichloro-1-hydroxyethyl) phosphonate
DDVP	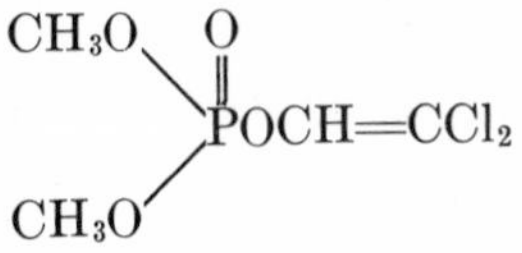	2,2-dichlorovinyl dimethyl phosphate

Table I.   Continued

Contact Insecticides

Common and Trade Names	Formula	Chemical Names
Imidan	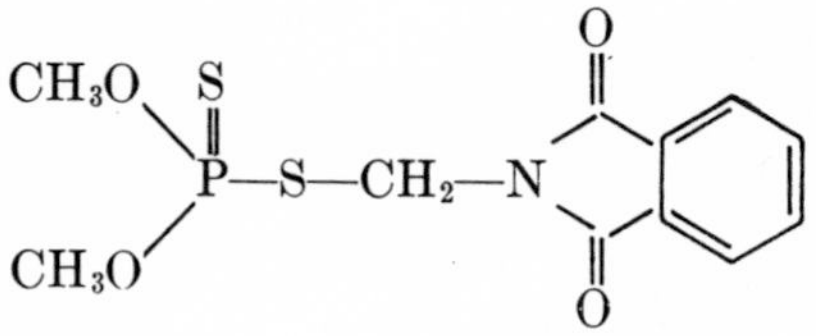	*O,O*-dimethyl *S*-phthalimido-methyl phos-phorodithioate
Dursban	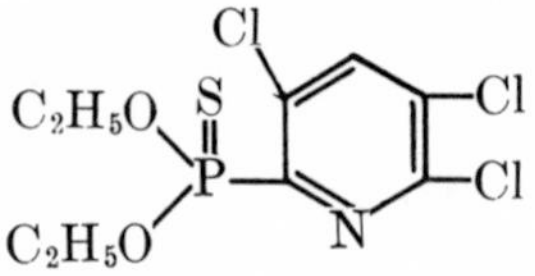	*O,O*-diethyl *O*-(3,5,6-trichloro-2-pyridyl) phosphorothioate

Systemic Insecticides

Common and Trade Names	Formula	Chemical Names
schradan	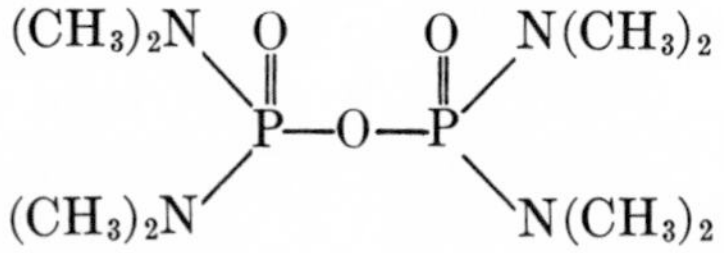	octamethylpyro-phosphoramide
Zytron	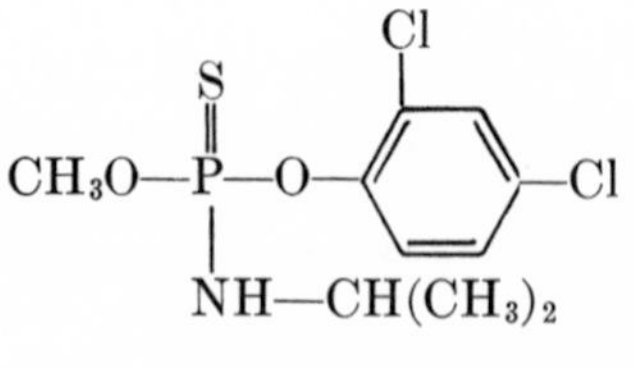	*O*-2,4-dichlorophenyl *O*-methyl isopropyl-phosphoramido-thioate
Phosdrin	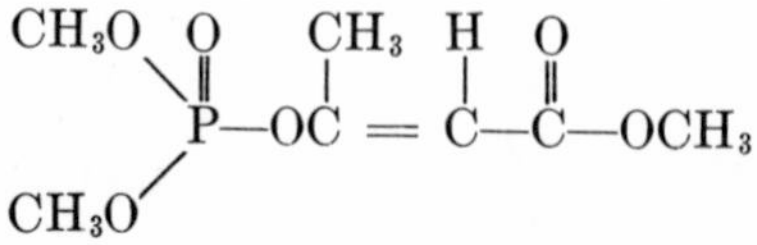	methyl 3-hydroxy-α-crotonate, dimethyl phosphate

Table I.   Continued

Systemic Insecticides

Common and Trade Names	Formula	Chemical Names
Amiton	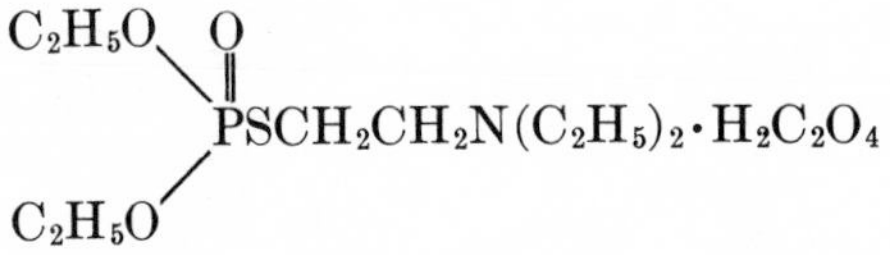 C_2H_5O, O ‖ $PSCH_2CH_2N(C_2H_5)_2 \cdot H_2C_2O_4$ C_2H_5O	2(diethoxyphos-phinylthio) ethyl-diethylammonium hydrogenoxalate
Demeton	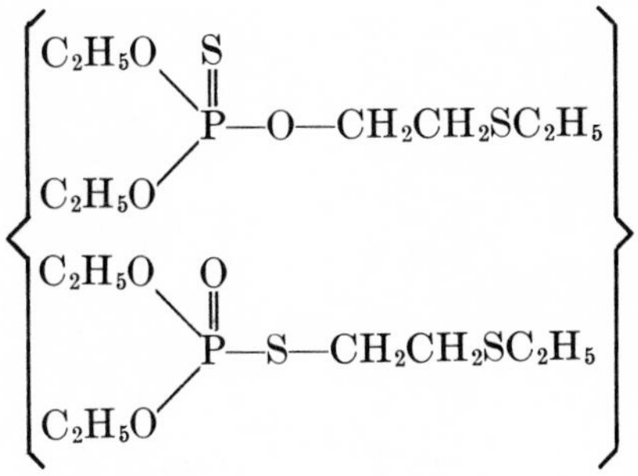 $\left\{ \begin{array}{l} C_2H_5O,\ S\ \text{‖}\ P{-}O{-}CH_2CH_2SC_2H_5\ \ C_2H_5O \\ C_2H_5O,\ O\ \text{‖}\ P{-}S{-}CH_2CH_2SC_2H_5\ \ C_2H_5O \end{array} \right\}$	mixture of O,O-diethyl S-(and O)-2-(ethylthio)ethyl phosphorothioates
Thimet	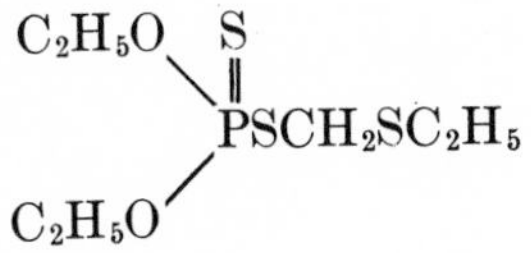 C_2H_5O, S ‖ $PSCH_2SC_2H_5$ C_2H_5O	O,O-diethyl S-ethyl-thiomethyl phos-phorodithioate
dimethoate	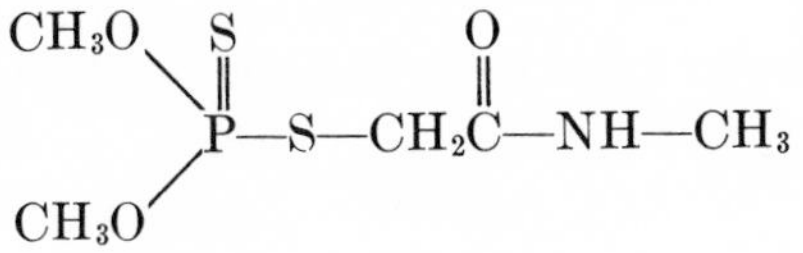 CH_3O, S ‖ O ‖ $P{-}S{-}CH_2C{-}NH{-}CH_3$ CH_3O	O,O-dimethyl S-(methyl-carbamoylmethyl) phosphorodithioate

Animal Systemics

Common and Trade Names	Formula	Chemical Names
dimethoate	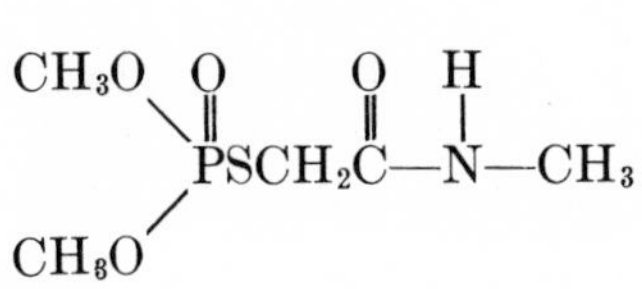 CH_3O, O ‖ O ‖ H $PSCH_2C{-}N{-}CH_3$ CH_3O	O,O-dimethyl S-(methyl-carbamoylmethyl) phosphoro-dithioate

Table I. Continued

Animal Systemics

Common and Trade Names	Formula	Chemical Names
Ruelene	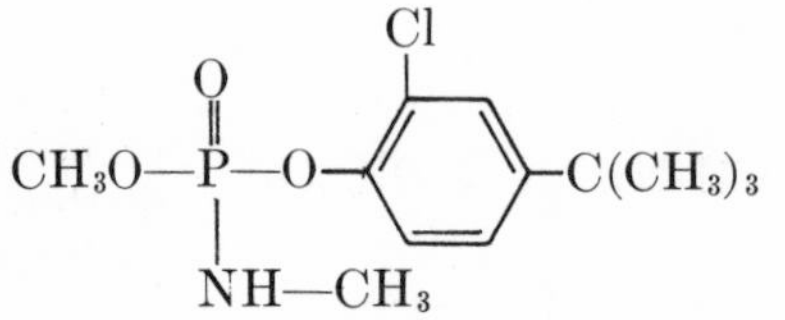	O-4-*tert*-butyl-2-chlorophenyl O-methyl methyl-phosphoramidate
Ronnel	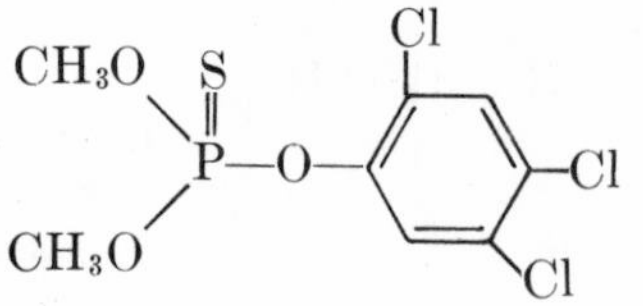	O,O-dimethyl O-2,4,5-trichloro-phenyl phos-phorothioate
Co-ral	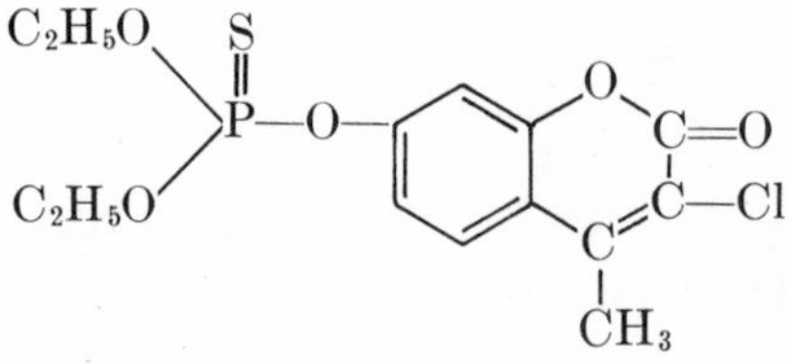	O-(3-chloro-4-methyl-2-oxo-2H-1-benzopyran-7-yl) O,O-diethyl phos-phorothioate

Contact Insecticides

Contact insecticides, as their name implies, kill when a compound actually touches an insect. Of the contact insecticides listed, both parathion and methyl parathion—besides being toxic to a wide range of insects such as cotton boll weevils, aphids, mites and caterpillars—are also quite lethal to warm-blooded animals. Nevertheless, they are still widely used, and because they are effective over a wide spectrum of insects, they are also economical.

In 1966, 19,444,000 lbs. of ethyl parathion and 35,862,000 lbs. of methyl parathion were produced. Average selling price was 75¢/lb. The price in 1971 was reduced to 40¢/lb.

EPN, a close structural relative of parathion, is toxic but not quite as dangerous as parathion. Parathion is used in Japan to control rice stem borers. However, since rice is grown in paddies, Japanese farmers also raise carp in these water areas. Unfortunately, parathion also kills the carp. EPN has a wide enough margin of safety so that it kills only the borers.

Courtesy Stauffer Chemical Co.

Differences in the quantity of corn root and resulting yield of corn from fields left untreated vs. those treated with Dyfonate. Dyfonate is a phosphorus insecticide which is used to combat corn root worms.

Sumithion has the same structure as methyl parathion except that it has a methyl (CH_3-) group substituted for a hydrogen in the meta position of the benzene ring in the p-nitrophenyl group:

methyl group
substituted for H.
↓

methyl parathion   Sumithion

This substitution, surprisingly, makes Sumithion a compound which has low toxicity to warm-blooded animals but is still able to kill a wide range of insects. In 1975, it was one of the insecticides used to control the spruce budworm disease. This disease is caused by the tiny larvae of the spruce budmoth which strips the needles from the tree for food. Millions of acres of spruce and balsam fir which are used for making paper are threatened by this disease.

Certain insecticides will render plants toxic to insects for many days. However, the U.S. Department of Agriculture forbids their use on fruits, for example, less than 30 days before harvest. At this point, other insecticides which kill pests and then are quickly hydrolyzed to harmless by-products can be used. For example, Phosdrin

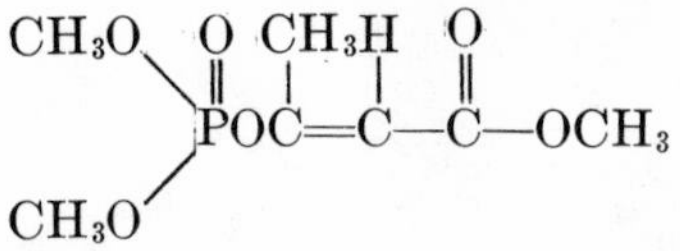

$$\begin{array}{c} CH_3O \\ \diagdown \\ POC{=}C{-}C{-}OCH_3 \\ \diagup \\ CH_3O \end{array}$$

can be applied to a field of lettuce to kill worms the night before harvest.

Malathion is used on a large scale to control garden insects. It is nontoxic to pets and children but because it has an unpleasant odor, it is not used indoors. Instead, compounds like Dipterex, Dursban, or Diazinon which are relatively free of bad odor, and are also quite non-toxic to warm-blooded animals, are used.

An accidentally discovered decomposition product of Dipterex has proved to be an interesting insecticide. In fact, many believe the toxicity of Dipterex is the result of its chemical decomposition to DDVP (dimethyl dichlorovinyl phosphate) *via* the following reaction:

$$(CH_3O)_2\overset{\displaystyle O}{\overset{\|}{P}}{-}\underset{\displaystyle H}{\overset{\displaystyle OH}{C}}Cl_3 \longrightarrow (CH_3O)_2\overset{\displaystyle O}{\overset{\|}{P}}OCH{=}CCl_2 + HCl$$

DDVP was also independently discovered and synthesized by the Perkow reaction, named after the German chemist who first published it in 1954. It involves the action of trimethyl phosphite on chloral and is illustrated by the following equation:

$$(CH_3O)_3P + Cl_3CCHO \longrightarrow (CH_3O)_2\overset{\displaystyle O}{\overset{\|}{P}}OCH = CCl_2 + CH_3Cl$$

trimethyl   chloral   DDVP
phosphite

A survey of the literature showed that this reaction had been dis-covered independently by several investigators both in Germany and in the United States within a period of a few months. This is a very good illustration of how people often simultaneously think along the same lines and come up with identical solutions.

One interesting recent development by Shell Chemical Co. is the household insecticide DDVP. When hung indoors, plastic strips impreg-nated with this insecticide release minute quantities of it into the atmos-phere. The vapor kills pests such as flies and mosquitoes. Since the insecticide vapor cannot be seen and yet knocks down flies quite effectively, it could be called the invisible fly swatter. The amount of insecticide released is of course insufficient to harm people, cats or dogs.

Systemic Insecticides

Systemic insecticides, as a class, are absorbed through the roots or leaves of a plant to render the sap toxic to sucking insects such as aphids without harming the plant. Contact insecticides sometimes kill both harmful and useful insects (such as bees) indiscriminately. They even kill insects that are the natural enemies of the harmful ones. Systemic insecticides do not harm the useful predators, which do not suck plant juices.

One of the first systemic insecticides discovered was schradan,

$$[(CH_3)_2N]_2\overset{\overset{O}{\|}}{P}O\overset{\overset{O}{\|}}{P}[N(CH_3)_2]_2.$$ The compound as synthesized by the chemist is not particularly toxic to insects or to warm-blooded animals. However, once it is absorbed into the plant, it is converted into a highly toxic chemical. This conversion may also be done by incubating it in slices of mammalian liver or even in ground-up lettuce leaves. The toxicity is increased a thousandfold by this incubation. Chemists and biologists believe that the toxic compound is an oxidation product of some enzyme system in the plants or animals and that the oxidation product may have the following structure:

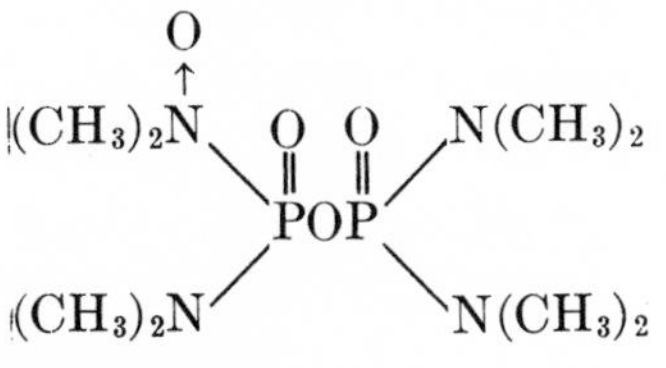

$$\begin{array}{ccc} & O & \\ & \uparrow & \\ (CH_3)_2N & O\;\;O & N(CH_3)_2 \\ & \diagdown \overset{\|}{}\overset{\|}{}\diagup & \\ & POP & \\ & \diagup \diagdown & \\ (CH_3)_2N & & N(CH_3)_2 \end{array}$$

oxidized schradan

If this is true, a similar increase in toxicity should be obtained by a chemical reaction. This was found to be the case. Schradan was oxidized with sodium hypochlorite to a highly toxic product, which however, is quite unstable. It seems best therefore to use just schradan and let the plant system do the converting to the toxic product.

The useful systemic insecticides are those that eventually decompose in the plants and leave no toxic residue. One systemic insecticide, Amiton, when injected into elm trees, effectively controls the beetles which are a major factor in spreading Dutch elm disease. Unfortunately Amiton's toxicity is so high and its stability is so great that if the treated elm tree

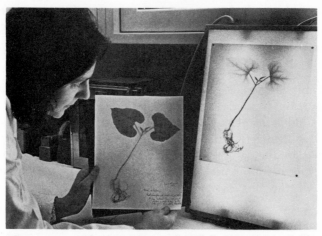

Plant grown in herbicide-treated soil is pressed on card at left, then photographed with a device called an autoradiograph—similar to an x-ray. Movement of the herbicide can then be traced from soil to roots, through the stem, and up into the leaves.

leaves were burned in the autumn, enough toxic smoke would be given off to endanger a whole community.

The research on organic phosphorus insecticides has advanced to the point where even animal systemics have been developed. Such compounds could be sprayed on or fed to the animals at a controlled dose that would kill insects feeding on their tissues but be harmless to the hosts.

The biggest development in this area involves the cattle grub. These parasites come from the eggs of heel flies which have laid their eggs on the hair of the cattle. The eggs hatch into maggots which tunnel under

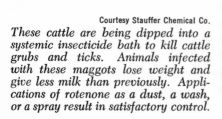

Courtesy Stauffer Chemical Co.

These cattle are being dipped into a systemic insecticide bath to kill cattle grubs and ticks. Animals infected with these maggots lose weight and give less milk than previously. Applications of rotenone as a dust, a wash, or a spray result in satisfactory control.

the skin of the cows and then bore through the animal's body in about six months. Finally, the maggot lodges in the back of the cattle and bites a hole through the skin for breathing. This causes a tumor-like inflammation about the size of a human thumb on the back of the animal. When the maggot has matured to about an inch long, it squeezes itself through the hole in the hide and drops to the ground. It will then develop into a fly and start the cycle over again by laying eggs on the cattle.

Cows infected with these maggots lose weight and give less milk. The maggots often lodge in the area from which prime steaks come, and hides with holes in them lose their value as premium leather. The systemic insecticides solve all these problems by killing the cattle grubs.

Research on organic phosphorus insecticides is still lively. The search continues for the wide spectrum insecticide that is toxic only to harmful insects and not to warm-blooded animals. The compound should be odorless, tasteless, and inexpensive. Its discovery and development represent a fascinating challenge to research chemists.

The research on organic phosphorus poisons is also a challenge to the biologist, the physiologist, and the entomologist. After all, we still do not have the complete picture of why certain compounds are effective while others are not.

INDEX

227